Air Canada
Air France
Air-India
Airlines
Braniff International (USA)
British Airways
British Caledonian
Delta (USA)
Eastern (USA)
El Al (Israel)
Lines
JAT (Yugoslavia)
KLM (Netherlands)
Laker (UK)
National (USA)
Northwest Orient (USA)
Olympic (Greece)
navia)
Saudia (Saudi Arabia)
Singapore Airlines
South African
United (USA)
US Air
Varig (Brazil)

AIRLINERS

indonesian airway

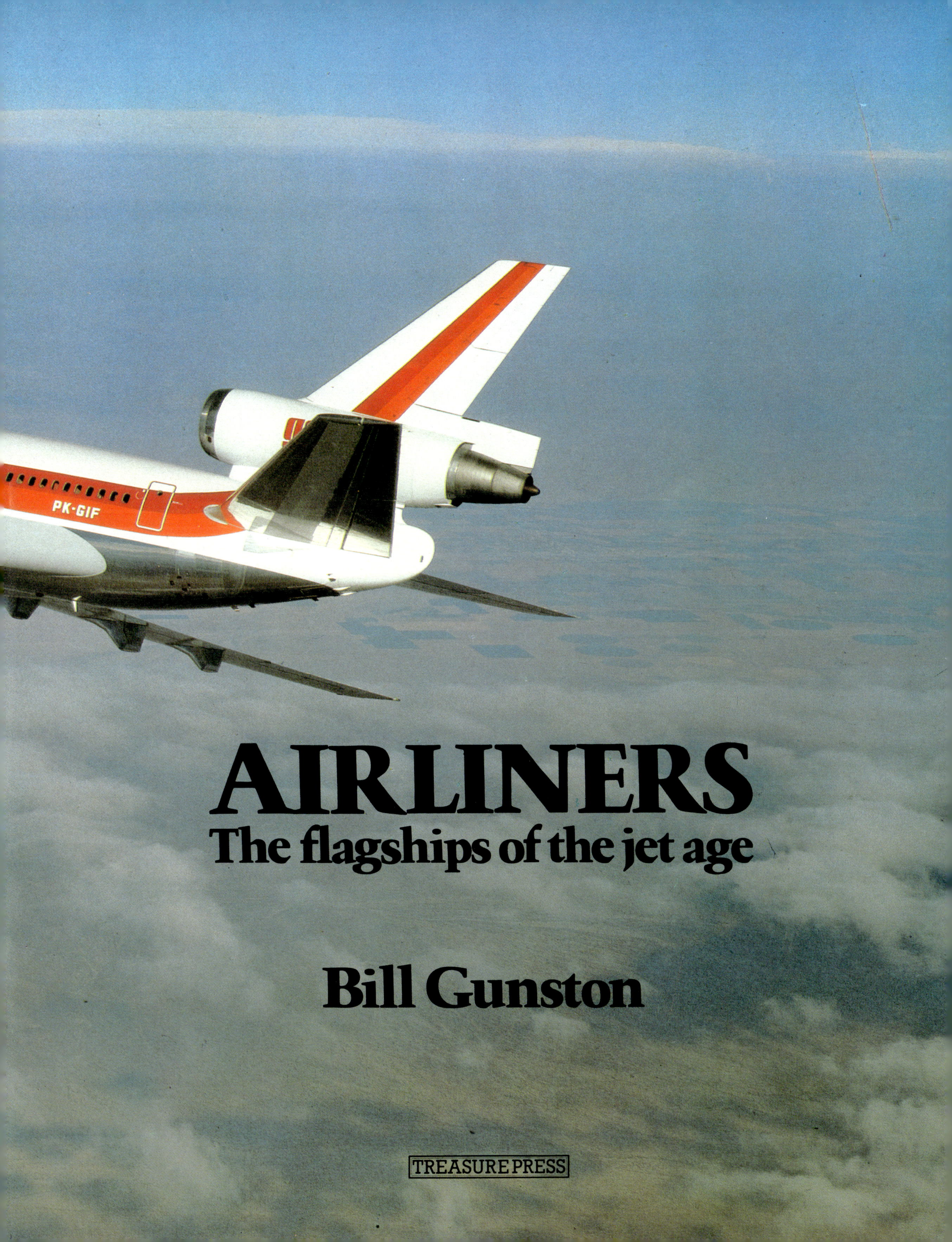

AIRLINERS

The flagships of the jet age

Bill Gunston

TREASURE PRESS

CAUTION

CONTENTS

First published in Great Britain by Octopus Books Ltd and distributed by Marks & Spencer Ltd

This edition published by Treasure Press 59 Grosvenor Street London W1

ISBN 0 907812 13 9

Printed in Hong Kong

PAGE 1: Boeing 747.
PAGES 2-3: McDonnell Douglas DC-10-30.
THIS PAGE: Flight deck of Concorde.

SE-
SE-BDC

INTRODUCTION

What is an airliner? The first aerial passengers were a sheep, a cock and a duck, which travelled by hot-air balloon on 19 September 1783. Humans followed just over two months later, but it was not until 22 June 1910 that the airline industry got off the ground – with airships. A former German general, Count von Zeppelin, not only made the airship a practical aerial vehicle but also started the first air transport company, Delag. By the time World War 1 broke out in August 1914 Delag's great silver ships had made 1,588 flights without harming any of its 33,732 passengers, most of whom came just to enjoy the amazing novelty of eating a sumptuous meal about a thousand metres above the ground.

On 1 January 1914 a small flying boat began fare-paying services across Tampa Bay, Florida, the passenger paying a surcharge if he or she weighed more than 91 kg (200 lb). But air travel remained virtually non-existent until the Armistice in November 1918. Then it was suddenly realized not only that aeroplanes would be useful in conveying

BELOW *The Douglas DC-6 was the most efficient of all piston-engined airliners, and probably the most commercially successful. First flown in 1946, 174 were sold by 1951. By this time the longer and more powerful DC-6A freighter and DC-6B passenger models were in production, adding 74 and 288 respectively. This DC-6 was one of 17 ordered by SAS of Scandinavia in 1946.*

ABOVE *Though in the 1920s engine reliability was (by modern standards) extremely poor, single-engined airliners were common. One of the best, and certainly the most aerodynamically efficient, the Lockheed Vega was also a celebrated record-breaker. One pilot said 'All you have to do is point her nose at a distant destination and there goes another record.' Vegas were wooden, and seated six passengers.*

military and political leaders to the various peace and disarmament conferences, and in organizing the Allied Control Commission that managed Germany and Austro-Hungary, but also that people and goods could be carried by air at a profit. At least, the idea was worth a try.

By 1919 Europe and North America were packed with unwanted aeroplanes and experienced pilots. To get things going, single-engined fighting aircraft and trainers, such as the British Bristol Fighter, Airco D.H.4 and Martinsyde F.4, the French Breguet XIV and the American Curtiss JN-4 'Jenny', were used with only simple modifications to carry a passenger or mail. Former bombers, such as the Handley Page O/400 and Vickers Vimy, and the French Farman Goliath, were modified to carry a few passengers in wicker chairs placed on the newly carpeted floor. Going even such a short distance as 300 km (190 miles) meant two to three hours of lurching and wallowing in an environment of vibration and indescribable noise, and passengers were advised to wrap up in a warm leather coat and bring barley sugar to suck. Some airlines issued hot-water bottles in winter.

Fundamentally, the most serious problem was that safety and reliability were by modern standards horrifying. It was commonplace to have to land in a field to ask the way, or – more often, and occasionally with tragic consequences – because of failure of an engine. Fog or mist were deadly, despite amazing skill and courage by the pilots. Rules of the air were written the hard way. In 1923, when there were only two flights between London and Paris each day, a British Handley Page and a French Goliath met nose-to-nose. This was because both pilots were flying directly above the same road for guidance, and they arrived over the village of Poix with unfortunate accuracy.

Such basically military machines were ill suited to this work. Their structures and engines were not up to it, and in any case they were totally uneconomic. Gradually, however, aircraft designers learned how to carry more passengers and baggage for less horsepower, and a tremendous contribution was made by the engine designers, who provided completely new engines designed for ten years or more of hard running. Eventually the airlines recognized that two aircraft companies could offer better products than the multitude of rivals. These were Fokker in Holland and Junkers in defeated Germany.

Fokker had built warplanes for the Germans, but in 1919 he smuggled trainloads of parts and material to a factory at Schiphol, Amsterdam. His airliners were monoplanes, with the wing placed above the cabin. He made the fuselage (body) out of steel tubes welded and bolted into a strong framework covered with fabric. The wing was wooden, and so thick that it did not need any bracing struts. During the 1920s he built nearly

500 airliners – several times the total for all British constructors – but he was outsold by Junkers, which delivered at least 350 of the F 13 alone.

Like all the aircraft of Dr Hugo Junkers the F13 was made entirely of metal, mainly of the new light alloy called Duralumin. To make the outer skin stiffer and stronger it was corrugated, the channels running fore and aft. The basic design of this unbraced low-wing monoplane was so outstanding that it outperformed all rivals from its first flight on 25 June 1919, and thanks to its tough metal skin some were still at work after World War 2! From this Junkers developed larger trimotor (three-engined) airliners, notably the Ju 52/3m.

Another great German constructor, Claude Dornier, pioneered the use of a Duralumin stressed skin that was not corrugated but smooth. In 1929 he flew a giant flying boat, the Do X, which was a monoplane and far larger than anything in use by the airlines. Powered by 12 engines in six tandem pairs above the rather stumpy wing, it once carried 169 people, including nine stowaways, but no airline had the nerve to buy it.

The largest aircraft anyone could sell in the 1920s carried about 18 passengers. Most airliners had just one row of seats on each side of the central aisle, and in some of them wicker seats were still used, but these were now fixed to the floor. A central aisle was required to allow passengers to reach their seats, but from 15 May 1930 it was also needed to allow a stewardess to attend to the passengers. A stewardess, who had to be a trained nurse as well as able to cope in all kinds of emergencies, was first carried by Boeing Air Transport (a forerunner of United, today the world's biggest airline apart from Aeroflot of the Soviet Union), aboard Boeing 80s, which were large three-engined biplanes. In the same month that stewardesses came on the scene Boeing flew the Monomail, a modern mailplane with a smooth stressed-skin wing into which the landing gears retracted in flight. The next step was a larger monoplane passenger airliner, the Boeing 247, which flew on 8 February 1933.

Often called the 'first modern airliner', the 247 was visually a break with the past. Sleek and streamlined, it had two cowled Wasp engines of 550 hp each, retractable landing gear and a smooth all-metal fuselage with enclosed cockpit seating two pilots side by side and five passenger seats on each side of a central aisle, each seat bolted to the floor beside a window. Extra refinements were rubber 'boot' de-icers along the leading edge of the wings and tail (described on page 27) to eliminate one of the greatest hazards of all-weather flying. United bought no fewer than 60 of these challenging machines, but the Douglas company responded with a bigger, faster and longer-ranged aircraft, the DC-1 (Douglas Commercial Type 1). Boeing came back with the 247D (D for developed), with better engine cowlings, three-blade propellers with variable pitch, so that the angular setting of the blades could be adjusted by the pilot for take-off or for cruising flight, improved pilot windshields sloping back instead of forward, and flaps on the wings.

To this competition between Boeing and Douglas was added a series of smaller high-speed transports by Lockheed, beginning with the wooden Vega and ultra-fast Orion, and in February 1934 switching to metal stressed skin with the Model 10 Electra, a speedy twin-engined machine with twin fins which established Lockheed as the third company on the US West Coast pre-eminent in the field of modern airliners (a class of aircraft called 'transports' in America). Since that time the United States, and these three companies in particular, have dominated the world airliner market. Fokker

BELOW Perhaps the most famous of all airliners, and by far the most widely used, the DC-3 is still found in almost every country. Douglas built 10,655, most of them as military transports during World War 2. This example formerly flew with Olympic Airways of Greece.

found it hard to change to all-metal stressed skin and faded from 1930. Junkers achieved great success with the last of its corrugated trimotors, the Ju 52/3m of 1932, but this could not compete with the Americans and in any case the company ceased operating at the end of World War 2.

By far the most important of the new breed of smooth-skinned monoplanes was the DC-3, descended from the DC-1 via the first production model, the DC-2 of 1934. The DC-2 had seven seats along each side of the cabin and 710 hp engines, and in a race from England to Australia in 1934 arrived with a full load of passengers and mail only a few hours behind the winner, which was a specially designed racer carrying no payload (commercial load) at all. To produce the DC-3 Douglas made the fuselage wider, so that on one side of the aisle there was room for double seats, allowing for 21 passengers instead of 14. With 1,000 hp engines and increased-span pointed wings, the DC-3 first flew on 17 December 1935, 32 years to the day after the first flight of an aeroplane by the Wright Brothers. Hundreds were soon sold and, including licence-production in Japan and the Soviet Union, some 13,300 were built. Most were military versions for World War 2, and thousands of these, with 1,200 hp engines, were sold cheaply after the war. The reason why so many are still flying is that the designers, Jack Northrop and Arthur Raymond, schemed the wing with several spars to carry the loads across the span. This happens to be a fatigue-proof structure; if one spar should eventually crack, the wing does not come off. In contrast British airliners designed in the decade after World War 2 almost all had single-spar wings and had to be withdrawn from use with fatigue problems.

ABOVE *First flown in 1926, the Ford Tri-Motor was basically a translation of the three-engined high-wing Fokker formula into metal, the structure and corrugated skin being aluminium. Over 200 of many versions were built. Popularly known as 'The Tin Goose', it was a mainstay of US airlines until 1939, and a few – such as this example at Las Vegas – survive to this day.*

Until World War 2 most airports were just small fields, often bumpy and rutted. The largest long-range aircraft could not safely use such fields, and for really long journeys airlines often used flying boats. These are aeroplanes with a fuselage whose underside is watertight and shaped to glide over the water on take-off and landing. In 1936 the main British airline, Imperial, planned its entire Empire operation on a new and very good fleet of flying boats which were a vast improvement over the outdated fabric-covered 145 km/h (90 mph) biplanes that it had used previously. Although a few flying boats continued into the 1950s, the world of air transport belonged to the landplane after 1945 because of the way the world had been covered with long paved runways during World War 2.

American domination in aircraft manufacturing, already apparent by the late 1930s, became almost complete in the post-war era. To some degree this was because American designers had more powerful engines, ranging up to 3,500 hp, and experience of all the new techniques such as pressurization. Pressurization means that, as the airliner climbs into thin air at high altitude, the pressure of the air is maintained by sealing the fuselage and pumping in fresh air, thus making the environment more comfortable for passengers.

Following the wartime DC-4 Skymaster – the first Douglas four-engined airliner – the chief post-war American machine was the much more powerful, pressurized DC-6, the most efficient airliner of its day. This family ended with the transatlantic-range DC-7C of the mid-1950s. Throughout this period, Douglas fought sales battles with Lockheed, whose Constellation (first flown in drab Army camouflage, on 9 January 1943) was the most powerful and most capable airliner of its era with the exception of the Boeing 377 Stratocruisers, double-deck monsters derived from the B-29 bomber but produced in small numbers. When Lockheed stretched their product into the 100-seat Super Constellation they had a vehicle which for several years sold successfully in the face of such new-technology British aircraft as the jet Comet and turboprop Viscount.

Today the jet airliner dominates the commercial aircraft scene, but there is still intense competition among short-haul turboprop aircraft. Although by 1960 the turboprop was widely thought to be made obsolete by the jet, with the soaring price of fuel it has come right back into popularity – along with such fuel-efficient jets as the new British Aerospace 146, which on estimated figures even beats the turboprops.

ABOVE *Britain's greatest chance to break the near-monopoly of the United States as builder of the world's airliners came at the end of World War 2 when the jet engine opened up new vistas of speed, altitude and comfort. This rare colour photograph was taken at Hatfield in July 1949 at the start of flight testing of the first jetliner, the de Havilland Comet. Fate decreed that six years later this potentially enormous programme should collapse because of a weakness in the fuselage.*

LEFT *In 1958 a new Comet, much enlarged and completely safe, went into service; one of these later Comets, a 4C, is seen in the background in Dan Air markings. But the airliner dominating this picture is one of the last piston-engined liners, a Lockheed Super Constellation. This is an L.1649A, the final super-long-range model of 1957.*

HOW AIRLINERS WORK

BELOW *A McDonnell Douglas DC-10 arriving at Los Angeles International Airport at sunset. A short/medium-range DC-10-10, without a centreline landing gear, it is similar to the aircraft that flew the very first DC-10 service between this airport and Chicago in August 1971.*

A railway enthusiast once complained that, since the demise of steam, locomotives all looked pretty much alike – and not all that different from the passenger coaches. One might have expected the same problem to diminish interest in airliners, as the majority are so streamlined there is apparently not much of interest to be seen externally. But in fact no two are quite the same, and even today we seem to be as far as ever from the one supposedly ideal and unbeatable design.

So, before looking under the skin at what is inside, it helps to take a brief look at the overall shape of modern airliners. Of course, the shape depends to some extent on what the aircraft is designed to do. Most designers agree the most difficult problem to solve is the cross-section of the airliner's fuselage; and, once this has been agreed, it is certainly the hardest part to alter.

The majority of passenger airliners in the immediate post-war era had circular cross-sections about 3 m (10 ft) wide, which allowed a double seat to be installed on each side of a central aisle. Unpressurized designs had flat vertical sides, and a few were double-deck machines with passengers above and cargo below. A few big aircraft, notably the Boeing Stratocruiser, had so-called double-bubble fuselages, looking like an inverted figure 8, with passengers in the big upper lobe and cargo below the floor. When the big 707 and DC-8 long-range jets were designed in 1952-4 a similar shape was chosen, but with the sides faired in by thin flat skins to reduce drag. But in 1966 Boeing planned the first of the so-called 'wide bodies', the 747. Thanks to a new breed of jet engine, the aircraft

ABOVE *Sold to the airlines after eight years of costly effort, the CFM 56 is a medium-sized high bypass ratio turbofan. The large fan, seen here at rest, was designed and made in France. Most of the slim but complicated core engine that drives the fan is American. The 97.9 kN (22,000 lb) thrust CFM 56-1B is in production to replace old engines on DC-8s.*

could be made very much larger, and, after studying various arrangements, it was decided to put all the passengers on one enormous deck running right to the nose. The flight deck was placed above, and below the floor were put all the masses of auxiliary systems as well as cargo and baggage holds and in many layouts the galleys and possibly even extra passengers. Some airlines put small lounges or extra seats behind the flight deck at the upper level, and in 1980 Swissair ordered the first of a new 747 version with the upper deck extended to the rear to accommodate another 40 passengers.

In the immediate post-war era all airliners had more or less the traditional shape of wing. Introduction of jet propulsion opened up a new realm of flight performance. It used to be possible to regard air as incompressible, but at speeds close to the speed of sound the pattern of flow around the aircraft alters. As the air flows around the aircraft it has to be speeded up locally (this is, in fact, why a wing gives lift: air is speeded up over the upper surface of the wing, and the pressure there falls, in effect sucking the aircraft upwards). The greater the width or thickness of the fuselage or wing, the greater the acceleration of the surrounding air. With jet aircraft it was soon found that parts of the surrounding airflow were reaching the speed of sound. This caused special waves, called shockwaves, to form in the air. Small shockwaves, as made by the tip of a whip, make a cracking noise. Large ones, as made by a stroke of lightning, cause a noise which at close range sounds like a bomb going off; at a distance they make the familiar rumble we call thunder.

Shockwaves and supersonic flow are inevitable for supersonic aircraft, but for subsonic (slower than sound) airliners they are undesirable because they greatly increase the drag and also can cause buffeting, vibration and flight-control problems. So designers found ways to eliminate them, by making wings thinner, sweeping them back like an arrowhead or by adopting new shapes, such as the triangular delta, making noses more pointed, sweeping back the tail and in general making aircraft longer and relatively slimmer. Most of the first generation of jetliners had quite broad wings swept back at 35° (sweepback is usually measured not at the leading edge but one-quarter of the way back across the wing, a location called 25% chord). One very fast aircraft, the giant 747, has more than 37° of sweepback, but since the 1960s designers have discovered how to make wings even more efficient.

The first airliner in service with the so-called supercritical wing is the European Airbus A300B. By 1983 it will be joined by the A310 and two new American types, the Boeing 757 and 767. A supercritical wing looks almost like a return to old shapes, in that it is deeper (thicker) and much less acutely swept. Sweepback is only 25° to 28°, which makes the wing perform better at low speeds and reduces problems of twisting and bending in flight. The wing avoids shockwaves by reducing the acceleration of the surrounding air. Instead of having a highly curved bulged region round the forward part of the upper surface, the supercritical wing has more of a bulge on the underside and a relatively flat top. As the wing can be deeper it is less severely stressed, and so can be of thinner and thus lighter materials. Inside there is room for more fuel, and not least of the advantages is that its aspect ratio (its slenderness seen from above) can be much greater. The longer and narrower the wing, the greater its aerodynamic efficiency.

However, there is an even more important reason for the greater fuel efficiency of modern airliners. In the 1950s designers had a choice between two types of engine: the traditional piston engine and propeller, and the turbojet. The latter made possible vastly higher speeds, but was less efficient in that on any given journey it burned much more fuel. Another drawback of the turbojet was that it was extremely noisy, though engineers found ways of easing this social problem at the expense of even greater inefficiency. A third alternative then was offered, the turboprop. This is a turbojet with extra turbines to take more of the energy out of the gas flow and put it into a propeller, driven via a reduction gear. This proved to be much quieter and to burn less fuel than the turbojet; but it restricted speed to only a little more than that possible with piston engines, and saddled airlines with the weight, cost and safety problems of propellers which the jet had eliminated, and which in most turboprop aircraft also inflicted noise and vibration on the occupants.

During the 1950s Rolls-Royce introduced a modified turbojet called the bypass jet, the first examples being the Conway, fitted to a few 707s and DC-8s and the VC10, and the Spey fitted to the BAC One-Eleven, Trident and F28. The bypass jet was designed to give better economy and reduced noise. This new jet was given an oversize low-pressure compressor at the front of the engine, handling more air than the engine could use. The excess was simply bypassed along a duct surrounding the hot core engine (the high-pressure compressor, combustion chamber and turbines) and discharged at the back, either separately or mixed with the hot jet from the core. It gave a modest improvement in fuel consumption, but the bypass ratio (the ratio of the cool airflow to that passing through the hot core) was nothing like enough. For the early Conways it was 0.3 and even for the Spey it is less than 1. By the 1960s engine designers could see that real boldness in choosing a bypass ratio of 5 or more would pay off in giving an engine with the economy and quietness of the turboprop with none of the latter's drawbacks.

It was the funding for a programme for a giant freighter for the US Air Force that made the modern high bypass ratio turbofan possible. General Electric won that contract, and soon offered a derived turbofan which is now flying in the DC-10, 747 and A300B, but the first in airline service was the rival JT9D engine from Pratt & Whitney around which Boeing designed the 747, the so-called Jumbo Jet. Later Rolls-Royce produced an even more advanced engine, the RB.211. All three companies realized that with a bypass ratio of at least 5 the engine can give immense thrust (generated almost entirely by a monster fan at the front) yet be economical on fuel and amazingly quiet. Such an engine might be thought unstreamlined, but of course its frontal area is deceptive because a hurricane of air – enough to fill a cathedral once a second – passes through it.

The jet age has certainly not increased the uniformity of airliner design as far as the arrangement of engines on or in the airframe is concerned. Piston-engined airliners invariably had one or two engines on each wing, and sometimes one on the nose. But with jets almost any arrangement is possible. The first jetliner, the Comet, had four engines buried inside the roots of the wings at the junction with the fuselage. The next, the Boeing 707, had four widely spaced engines hung in external pods below and ahead of the leading edge of the swept wing. Following this, the Caravelle had one engine hung externally on either side of the rear fuselage. The Trident introduced the trijet formula,

BELOW *Airliners undergoing major overhaul are often towed into a surrounding dock giving all-over access. This view of a British Airways 747 tail shows alternative ways of reaching particular parts: a powered telescopic cab hung from the roof and an elevating mobile stairway.*

with the centre engine inside the tail of the fuselage and the others close on each side. The TriStar has two of the three engines hung on the wings, and the DC-10 offers a variation in which the centre engine is in a straight-through duct above the rear fuselage, on top of which is mounted the vertical tail.

To help fast jets slow down on landing, experiments began in the early 1950s to develop a system for reversing the thrust of turbojets and this was first used on the Comet 4 and early 707s. These reversers featured upper and lower internal 'clamshell' shutters inside the jetpipe. When the pilot selected the reverse position the shutters closed the jetpipe and opened large apertures on the left and right fitted with a series of curved vanes which deflected the jet diagonally forwards to help pull the aircraft up on a wet or icy runway. Today most turbojet and bypass-turbojet reversers are of the target or bucket type. Large external sections of the engine pod, usually above and below the jetpipe, swing out and to the rear on parallel arms to meet downstream of the nozzle.

Today engine management in the latest airliners is computerised. The flight crew, which comprises either two or three pilots, merely tell the computer what they want the aircraft to do and the engine (or rather energy) control system then flies the aircraft so that energy consumption is minimized, thus saving fuel and noise.

With aircraft such as the 707, with engines spread along the wing, failure of an outer engine causes a major asymmetric problem. Unbalanced thrust of the outer engine still operating on the other side swings the aircraft to the left or right, and the pilot flying the aircraft has instantly to apply a powerful force on his rudder pedals to hold the aircraft straight and make it climb away with wings level. Lack of asymmetric problems is one of the advantages of grouping all engines at the rear, where, if one fails, the aircraft merely loses thrust and has little tendency to swing or rotate to left or right (a movement known as yawing).

This leads into the vital area of flight control. Traditionally aeroplanes (heavier-than-air machines with fixed wings) are controlled about their three axes by three sets of controls. For control about the longitudinal axis, in what is called roll, the pilot uses hinged surfaces on the outer wing trailing-edge called ailerons. These are worked by turning the control wheel or yoke to left or right. To make the aircraft climb or dive (rotate about the lateral axis) he uses elevators, hinged to the trailing edge of the tailplane (the horizontal tail, called the stabilizer in America), worked by pulling the wheel or yoke back to climb or pushing it forward to dive. To control about the vertical axis, in yaw, he uses the rudder, hinged to the back of the fin, worked by left and right pedals; he pushes the left pedal to make the nose yaw to the left, or indeed to correct the yaw caused by a non-functioning right engine.

Modern jetliners often have modified arrangements. Instead of, or as well as, ailerons they may have spoilers for roll control. These are flat surfaces hinged along the top of the wing, and often they can all be opened together to increase drag and reduce lift and help the aircraft descend very rapidly without going too fast. After landing, some spoiler sections, called lift dumpers, may flick open automatically to destroy wing lift, put the weight firmly on the wheels and thus increase the power of the wheelbrakes. Many jetliners have two sets of ailerons, one pair inboard (and available at all speeds) and a normal pair near the tips which may be locked at high speeds. The traditional idea of a fixed tailplane and hinged elevators is also uncommon on modern jets, which either have a

BELOW *Assembled at Amsterdam, Fokker's F28 Fellowship is a short-haul jet especially well adapted to small airports. This touchdown picture shows the non-skid brakes in action on a dirt runway, the tail airbrakes open, and all the spoilers and lift-dumpers raised on the wings.*

hinged tailplane and an elevator used to increase its power by in effect increasing camber (the curvature of the tail surface) or have just one big surface driven directly by the pilot, with no elevators at all.

Of course the power required to drive the control surfaces on a big jet is many times more than a pilot could exert. For 30 years most large airliners have had powered flight controls, and the control systems of modern aircraft can be amazing in their complexity. They must obey the pilot instantly, with a force of many tonnes, yet with the delicacy of a fine watch, while allowing for major changes in aircraft dimensions and shape caused by violent changes in temperature or severe flight loads. The surfaces themselves are invariably driven hydraulically, like most other large movable items such as the flaps, slats and landing gear. Some airliners increase safety by driving each surface by three power units, or by two twin pairs, each of which helps to overcome any incorrect push applied by a failed unit.

What about the slats and flaps? These are not flight-control surfaces but movable surfaces of great power that, by changing the cross-section or 'profile' of the wing, greatly increase its lift at low speeds. Slats were invented in 1919. They are long curved strips, rather like a large piece of guttering, which when stowed fit flush along the leading (front) edge of the wing. At low speeds, when the aircraft has to fly nose-up to stay in the air, the slats may be pulled open automatically by the aerodynamic lift of the reduced air pressure above them, but in modern jetliners they are usually selected by the pilot and forced open by hydraulic power. When open they speed up the airflow over the leading edge and force the air to flow back across the wing, thus continuing to generate lift. Without them the flow would break down and the wing would 'stall', the aircraft dropping like a brick. A modern variation is the droop leading edge, or droop snoot, in which the leading edge of the wing itself can be hinged down for better lift at low speeds. Another is the Krüger flap, a leading-edge flap which swings down and forwards from below the leading edge to turn the wing from a sharp-edged high-speed profile into a blunt surface that lifts well at low speeds.

Main flaps were first used on airliners 50 years ago. Early ones were merely hinged sections of wing trailing edge, or (the 'split' type) large plates hinged down below the trailing edge, in each case to increase lift and/or drag. Today they are large and complex structures, driven by hydraulic motors generating hundreds of horsepower and positioned with great precision. A popular type is the Fowler flap which rides on small carriages running out behind the wing on rails whose fairings are often prominent projections; for example, they can be seen along the rear underside of the wing of a 747, TriStar or Airbus, while the DC-10 has deeper but

ABOVE *With the 727 trijet Boeing created a high-speed airliner that could be used from typical runways at major airports in the early 1960s. The result has become the best-selling jetliner, with orders approaching 2,000. This British (Dan-Air) 727 has its slats and triple-slotted flaps fully extended. The main wing box, on which the registration is painted, forms the integral fuel tankage.*

FLAP

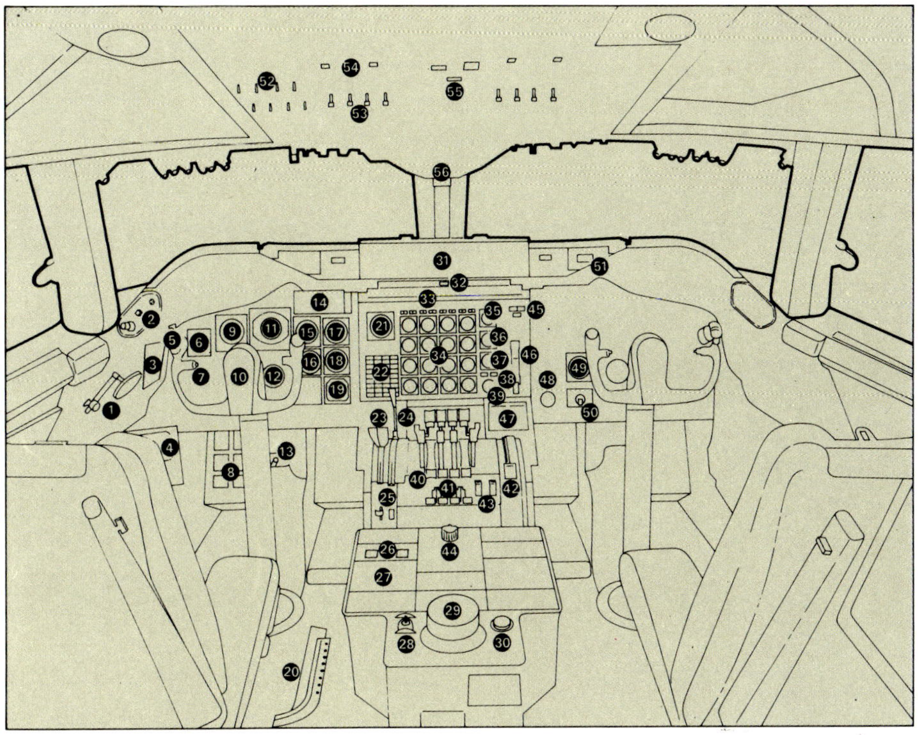

Flight deck of a Boeing 747

1 Nose gear tiller
2 Panel, map and clock lights
3 Chart holder
4 Weather radar scope
5 Tailplane trim switch
6 Clock
7 Autopilot disengage switch
8 Rudder pedals with wheel brakes
9 Mach and airspeed indicator
10 Pilot's control yoke
11 Attitude director
12 Flight compass director
13 Rudder pedal adjustment
14 Flight-system mode annunciator with navigation marker-beacon lights (on right)
15 Altimeter
16 Vertical speed indicator
17 Radar altimeter
18 Automatic direction finder display
19 Indicator for control-surface positions
20 Seat track
21 Standby attitude director
22 Main warning panel
23 Manual tailplane trim lever
24 Speed-brake handle
25 Parking brake latch with warning light alongside
26 Automatic direction finder controls
27 VOR/ILS/DME navigation controls
28 Roll trim
29 Rudder trim
30 Warning horn silencer
31 Autopilot control panel
32 Urgent warning light
33 Thrust reverser indicator lights
34 Instruments for the four engines arranged vertically
35 Air temperature indicator
36 Flap position indicator (outboard)
37 Flap position indicator (inboard)
38 Leading edge flap lights
39 True airspeed indicator
40 Throttle levers
41 Engine start levers
42 Flap selector
43 Tailplane trim cutout switches
44 Pitch and turn commands for autopilot
45 Landing gear indicators
46 Landing gear control handle
47 Computer keyboard
48 Brake hydraulic indicator
49 Radio compass
50 Computer selector switch
51 Aircraft registration number or letters
52 Flight controls and hydraulic power
53 Engine start panels
54 Engine fire extinguisher panels
55 Fire warnings and emergency lighting
56 Standby compass

NB: Co-pilot's instruments on right of cockpit are the same as the captain's.

ABOVE *At the largest airports, hydrant refuelling speeds the turn-round of jetliners which, in the case of this 747 of British Airways, may need 205,000 litres (45,000 Imperial or 54,000 US gallons). The kerosene-type fuel is stored in underground tanks and pumped aboard by trucks carrying only hosereels, pumps and filters. An elevating platform enables the fuelling company crew to lock the hose coupling into its socket, in this case under the wing outboard of No 3 engine.*

shorter projections that house external hinges for flaps of the double-slotted type. In all cases, the initial motion carries the flap out behind the wing to increase the wing's area and, by adding a little camber (curvature), greatly increase the lift. Further motion not only pulls the flap out even more but also rotates it trailing-edge downwards to give a very large increase in drag, without spoiling the extra lift, to slow the aircraft after landing. So on an icy runway a jetliner can have large flaps, spoilers, lift dumpers, non-skid wheelbrakes, reversers and, in a few cases such as the Il-62, a braking parachute! Virtually all jetliners have multi-wheel main landing gear to spread the load over the largest possible area, and steerable nosewheels for accurate taxiing.

Except for short local flights by trainers and light private machines all flights have to be planned in advance; a 'flight plan' is a written document stating the aircraft, crew, origin and planned destination, alternates (strictly the word should be 'alternatives') to which the aircraft would divert if a landing at the original destination should for any reason be impossible, and precise details of scheduled times and many other items. In the case of airline flights the captain is not necessarily free to choose his route or cruising height(s). He has to maintain specified minimum separation between other aircraft in terms of distances and flight-times and hold the exact assigned Flight Level (height) which cannot be assigned to traffic moving in any other direction.

Traditionally the take-off involves simply opening the engine throttles to full power, but today the latest airliners have computer control which draws from the engines exactly the power needed. By avoiding the use of full power except on rare occasions the engine life is greatly extended. Already, with what is called on-condition maintenance (engines are left alone until one of many sensitive devices says trouble is brewing), modern turbofan engines may cross the Atlantic 100 times without any attention except an occasional inspection. On take-off the aircraft is 'rotated' nose-up at an exact speed that depends on the aircraft weight and the airfield height and atmospheric temperature, and climbs in a way that is known to be safe even if an engine were to fail at the worst possible moment. At the top of the steep initial climb the aircraft is rotated back into a much shallower climb and the engines cut right back to reduce noise as the aircraft crosses the airport boundary and traverses built-up areas.

Once in the air, how do airliners navigate? Since 1957 the most common aid has been VOR/DME, very high frequency omnidirectional range with distance measuring equipment. VOR/DME stations cover the entire globe, and when a pilot tunes in to one he reads off an instrument the heading to steer to reach that station. Unfortunately, this method channels traffic along congested narrow pathways called Airways (unless the pilot has a recently developed and low-cost method called G-Nav which enables him to take the quickest route).

There are many other navigation assisters, such as gyromagnetic compasses of great accuracy, nose radar which not only warns of storms but can picture the ground, and a radar-like method called Doppler which – by using a well-known physical effect which makes the pitch of a train whistle appear to fall as the train passes the observer – can give an exact measure of speed over the ground and lateral drift, which of course both depend on the wind. But today most large airliners base their automated navigation on inertial systems, which were pioneered with bombers and missiles and are totally self-contained. An INS (inertial navigation system) usually comprises a small platform that is held exactly parallel to the local horizontal direction and on which are mounted accelerometers that measure acceleration fore/aft, left/right and up/down. They are the most accurate mechanisms yet created by man. All the pilot has to do is check that the INS knows the start position, which is the particular gate at the airport where the aircraft is parked. Then, as the airliner taxies out to the runway, the INS meticulously records the imposed accelerations. From these it works out the speed at each moment, and the direction. By 'integrating' these measures throughout the flight the INS indicates the exact position.

An INS is one of many types of R-nav, or area-navigation system, so-called because they free aircraft from the confines of VOR/DME and allow them to fly just where they want. Of course this could complicate the ATC (air traffic control) problem, but as nearly all controllers in high-traffic areas watch the traffic on radar they can always maintain acceptable standards of separation,

which is the jargon for saying that they prevent mid-air collisions. Each aircraft leaves a small bright mark, or 'blip', on the radar display which the controller has to identify. Modern surveillance radars are linked with an SSR (secondary surveillance radar) which 'interrogates' each aircraft with a complex coded sequence of electronic pulses. Airliners carry a transponder (transmitter responder) which, when triggered by the interrogating pulse, instantly sends back a stream of answering pulses giving the flight's identity.

Near the destination the en route systems hand over control of the flight to an arrivals or approach controller. Often the captain may be advised to hold in a stack because the airport is congested. He has to fly a precise racetrack pattern at an assigned flight level (FL), after a while being cleared to successively lower FLs until with much relief the flight is permitted to leave the bottom of the stack and fly to intercept the extended runway centreline perhaps 15 km or more from the airport. Along the centreline of the runway the ILS (instrument landing system) sends out two sets of radio beams, both comprising two sets of coded radio signals at different frequencies. One beam, the localizer, gives left/right directional guidance, so that even in a strong crosswind the captain can steer accurately on to the centreline of the runway. The other beam, the G/S (glide-slope), is pointed up into the sky at a shallow angle to give vertical guidance. The pilot has only to keep the two pivoted needles of his ILS indicator crossed in the centre at 90° to arrive at the correct touchdown point.

After landing, the pilot flying the aircraft steers with a small wheel or tiller that controls the angle of the nosewheel(s), while retracting the flaps and spoilers and restoring the reversed engines to the forwards-thrust regime. At some airports nose-in parking is adopted: the airliner gently rolls straight in at 90° to the terminal, or to a projecting 'finger', the exact stopping place being shown by any of various kinds of electronically displayed indicators that know the type of aircraft and place it so that the terminal airbridge is exactly opposite the passenger door.

Some airliners, the pioneers being the British VC10 and Trident, can land automatically. Their systems not only fly the aircraft down the approach to the runway without the flight crew being able to see the ground, but sub-systems automatically command the flare (the pulling back on the controls to arrest the rate of descent), line the aircraft up with the runway (in a crosswind it will have been 'crabbing' at an angle), kick off any residual sideways drift, close the throttle, put the aircraft on the runway and automatically deploy the lift dumpers and reversers. Automatic roll-out guidance steers the aircraft to the parking gate. The captain then goes through customs, gets into his car and then cannot drive home because of fog.

ABOVE *One of the ways of arranging a major airport is to build projections called fingers to carry passengers between the terminal and the aircraft gates (parking stands). Here at London Gatwick a single finger connects up to 18 long-haul jets (most of which dwarf the 707s seen on the left, which 20 years ago were frightening airports because of their size).*

FRONTIER

ON BOARD A MODERN JET

Airliners are one of the very few of man's inventions to have enjoyed the happy association of traffic and price. More traffic – more passengers, cargo and mail – means that the cost of moving each item falls, so fares can be steady or even reduced. If fares are reduced, more people and goods take to the air. Except perhaps for modern microelectronic gadgets, none of man's creations has grown so fast as air travel, and the key to this remarkable growth is the increasing size, efficiency and safety of the modern airliner.

Today the 737 is the baby of the Boeing Commercial Airplane Company, the No 1 builder of airliners. A typical short-haul jet, it seats from 100 to about 130 passengers in all-passenger configuration (this qualification has to be added, because like many modern jets it can have all or part of its main deck devoted to cargo). Yet in the years immediately following World War 2, only some 30 years ago, an equivalent airliner was the DC-3. This carried just 21 passengers. Going back to the early 1930s we find that a 15-seater was considered large; most of the aircraft on the world's airways had fewer than ten passenger seats, and on many flights there were only one or two passengers. Worse, it was not uncommon for airliners that operated to a strict schedule often to fly with nobody on board but the pilot!

Of course, there are a few older people who think that the era of mass travel is a retrograde step. They would like the clock put back to the days when only a few exclusive people flew – for example, in stately flying boats, which cruised at not much

BELOW *Parked at Salt Lake City, where the Mormon settlers arrived in covered wagons, a DC-9-30 of Hughes Airwest awaits its next load of passengers. Only a few years ago this 115-seater would have been judged a giant; today it is routine, and there are three bigger models even in the DC-9 family!*

ABOVE *When Airbus Industrie planned the A300B care was taken to make this wide-body twin a good cargo carrier. Even with all the 275-odd seats filled there is room for more than 14 long tons of cargo, packed in jumbo-size containers and pallets. Here cargo goes under the floor of a German A300B2 while a commissary truck services the galleys.*

more than 200 km/h (125 mph) with a mere handful of interesting people sharing two decks with a purser, stewards, silver cutlery and potted palms. Airships even boasted grand staircases and, bolted to the floor through a thick-pile carpet, a grand piano. Perhaps the airline industry grew up when it recognized that its job was not to emulate the bygone era of passenger ships but to carry the people of the entire world, and their goods, fast and cheaply. Today's load of 500 sunburned people happily going home to fog and rain from Majorca prefer cheap holidays to potted palms and grand pianos.

So today every cubic centimetre of an airliner is carefully allotted to a particular task. Thousands of kilometres of electric cables, fine wiring, hydraulic pipes, giant hot or cold air ducts, fuel pipes and oxygen lines fill spaces that extend to every part of the aircraft, for all the world like the blood vessels and nerves of a human. The passengers see nothing of these vital arteries, though they may occasionally hear hisses, clunks and thumps caused by a thousand small devices or such major units as the main landing gear whose groups of wheels are normally retracted directly under the cabin floor. Just a part of the clever design of a modern jet is that, unlike the wallcovering in a modern home, virtually the entire interior furnishing trim is arranged to be unclipped or unscrewed in a matter of minutes. This is partly to give engineers quick access to the mass of 'nerves and arteries' but also so that the furnishing may be easily replaced when it begins to become worn.

Wear and tear can be tough on a modern jet, which may make six trips a day. Vandalism is almost unknown in aviation but the hard work and long life of modern airliners means that the furnishing trim may have to be replaced several times. In the 1930s it was rare for an airliner to stay in business longer than five years, whereas today few jets are retired within 20 years and most look like going on for 30 or 40. And, whereas in the pre-1939 era a utilization of 400 flight hours a year (just over an hour a day on average) was considered excellent, today airline bosses would complain if a long-hauler did not clock up ten times this amount. Even in short-haul operation, where – with up to ten flights a day by each aircraft – much of the time has to be spent loading and unloading, a utilization of 2,500 hours a year is typical of good operators. All this is reflected in the hard work of ground staff who not only clean the interior during a ten-minute 'turnaround' between flights, and service the toilets, drinking water, coffee (and on occasion complete galley loads of meal trays or duty-free stocks of liquor and cigarettes) and slip reading material into seat pockets, and check that countless other small items are exactly right, but also from time to time strip out the curtains, wall trims, seats, partitions and other scuffed or 'tired' items

and put in new ones ready for the next flight.

Top designers are invariably called in to advise on interiors. Each airline likes to create and present a strong 'house' image, with matching colours, logo and materials. Of course, this also affects the outside of the aircraft, where epoxy-based paints that stand up to bombardment by rain and hail at close to (or, on Concorde, more than twice) the speed of sound form a tough skin that never fades in brilliant sunshine or intense cold at high altitude. Quite often airlines lease aircraft from others, and then their own name and logo has to be stripped on the outside in a gigantic series of peel-off strips which stay on in the hurricane slipstream and then peel off when the lease is over.

Many airlines like to add extra evidence of their identity round doors, so that when cameramen film the arrival of famous people the airline gets free publicity. But what the public would seldom even notice is the doors themselves. In early airliners the doors were flimsy structures of steel tube, wood or aluminium, often covered with fabric. The new breed of all-metal stressed-skin airliners in the 1930s standardized on doors with light-alloy skins like the rest of the aircraft, but they could still be carried by a man. The big change came with pressurization. Today the door has to be an airtight seal able to withstand the interior cabin pressure, which can try to open it with a force of several tonnes. The giant doors of convertible passenger/cargo aircraft may have to remain sealed against an in-flight load of 50 to 60 tonnes. Actually effecting the seal is relatively simple: a flat rubber pipe around the door is inflated in flight to seal the gap in such a way that the interior pressure forces it against the door frame. In the earliest pressurized airliners the doors were held shut by one or more latches; if they broke, the door would burst open, the resulting 'explosive decompres-

LEFT *The Lockheed L-1011 TriStar wide-body trijet has three cargo doors and three passenger doors on the right side, and three more passenger doors on the left. Here passengers disembark past an open cargo door down one of the TriStar's unusual options: built-in folding stairways. These make the aircraft independent of stairs or jetties at the airport, but reduce the maximum payload carried.*

sion' hurling passengers and other loose objects out. Now this cannot happen, for the simple reason that the doors are larger than the door opening, and cabin pressure just forces them to seal more tightly. To open them the cabin crew flick a lever and the door either opens inwards and slides up round the inside of the roof, or sections at top and bottom retract and the shortened door swings out on large arms to lie along the outside of the fuselage.

For any ground emergency large escape chutes are carried either in compartments below the main doors or on the backs of the doors themselves. They are inflated from gas bottles and in seconds the whole load of passengers can slide to safety, bodily helped out by the cabin crew. Also, there are always extra doors and windows constructed as emergency exits. In addition there are extremely comprehensive fire-protection systems triggered by sensors for excessive temperature, smoke or other symptoms. Ice forming in the environment of supercooled water droplets in the cold clouds (such a deadly enemy in the early days of aviation) is virtually eliminated by various systems relying on hot-air heating, electric heating or, in the case of slower aircraft, pulsating rubber 'boots' along the leading edges. These boots, which are prominent black strips, are very flexible and pumped up and down in sections by air pressure, breaking up the thin skin of ice as fast as it forms. There are many other provisions for emergencies, such as drop-down oxygen masks which dangle in front of each passenger in the event of sudden loss of cabin pressure, and large inflatable dinghies, packed with survival gear, which are released in the event of a ditching. The latter is an almost unheard-of event; it would make headlines if it were to occur.

In a ditching the sharp deceleration could reach as much as 6 g (six times the acceleration of Earth gravity). Many experts think this can best be reacted by seats facing to the rear, and this kind of seating is common in military transport aircraft. However, according to airlines most passengers prefer to face forward, so they have to rely on their single lapstrap. The flight crew have full seat harness passing over the shoulders. All airline seats are stressed to resist at least 9 g, and flight-crew seats are usually even stronger. Most have frames of steel tubing, with extremely comfortable seat and back, the latter being adjustable and incorporating a table for the passenger behind. Armrests usually contain ashtrays, a recline button, and occasionally the passenger service panel with controls for fresh air, lights, cabin-service call and entertainment such as radio or film soundtrack; more often these complex panels are in the underside of the overhead stowage. The latter was formerly an open shelf like a luggage rack in a train. Today the newest airliners all have large bins for each passenger with latchable doors, in which they can put anything that will fit.

Almost all travel on most routes is one-class, generally called 'tourist'; but on many trunk routes special first-class service is offered at a substantially higher price. Seats are fractionally larger, aisles are wider and the accommodation is provided at the very front of the aircraft behind the flight deck (right in the nose, in the case of a 747) where noise is at a minimum. Another difference is that everything possible is done in first-class to offer a lavish meal, any amount of free drink and instant service, whereas the common herd have to request attention and get only a good cafeteria-style meal.

Invariably, the cabin crew of a big jet are busy and even harassed throughout the main portion of each flight. Certainly everything possible is done to relieve their work-load, and the old-time practice of actually preparing meals in flight is unthinkable. Airline catering is big business, specialist firms at some airports having contracts calling for as many as 100,000 meals a week plus many supporting services such as disposable cutlery and napkins. Precooked or frozen meals, often of varied style and with particular national or religious dishes included, are reheated on board the aircraft in microwave ovens in seconds and served to passengers at the rate of one a second. The wide-body jets, seating 200 to 500, have as many as six beautifully designed galleys and some aircraft, such as the TriStar, have galleys at the lower-deck level with lifts to shuttle large meal-tray and bar-service trolleys up to the main level. Meanwhile, other cabin staff are helping mothers with babies, making special arrangements with the destination for a passenger in a wheelchair, searching for a place for guitars and harpoon guns, finding a common language in which to speak to someone from an unusual country, and answering unbelievable questions – for example, that there is another wing and engine on the other side of the aircraft!

LEFT *Galleys are big business, the contract to equip a fleet of large airliners being a much sought after concession. The galleys of a TriStar, for example, can handle the storage and preparation for serving of 600 full-course meals. Sometimes meal trays are brought on board already packed into service carts.*

ABOVE *When the 747 went into service in 1970 passengers loved the spaciousness of this pioneer wide-body. Today this comfortable travel, with no passenger more than one seat from an aisle, has been extended to shorter inter-city routes by the Airbus A300B, as shown here. At present the only way to beat wide-body travel is by paying more than twice as much and going first-class (see below).*

OPPOSITE *First-class passengers enjoy not only more space but also constant attention from what the rest of the travellers may think an unfair proportion of the cabin crew. In some wide-bodies, such as Saudia 747s and TriStars, the single seats can be moved on tracks and pivoted.*

LONG-HAUL JETS

Long-haul is the airline way of saying long-range; and range is the aviation word for the distance aircraft can safely fly after taking off with a particular amount of fuel. Just to make the concept of 'range' clearer, but unfortunately more complicated, all airliners not only have a fixed limit to the quantity of fuel they can carry but they also have a fixed limit to the weight at which they may take off. Almost always, if they have a full payload of passengers, cargo and baggage, they cannot also take on a full load of fuel. Conversely, if they fill up all the fuel tanks completely, they cannot accept a maximum payload.

Airline people often plot the payload of an airliner against the range. Usually the graph has a straight top line out to a certain distance, say 2,000km (1,250 miles), at which point the payload begins to fall off with range until it becomes zero at some greater distance, perhaps 4,000km (2,500 miles). In other words, the aircraft can just fly 4,000km with maximum fuel and no passengers or cargo; but that is no good to an airline. So what is the range? Usually it has to be related to payload, and often the figure chosen is the 'knee' of the graph where the payload first begins to fall. This is called the maximum-payload range, in other words the greatest distance the airliner can fly with a full load of passengers and cargo. To show how big the difference can be between this range and that with no payload, a Boeing 747-200F, the 'freighter Jumbo', can fly 5,920km (3,680 miles) with a full load of 91 tonnes of cargo; but if the cargo was off-loaded and the tanks filled to the brim it could safely fly 12,950km (8,050 miles).

Obviously the 747 is a long-hauler. But where is the dividing line? This is not easy to fix. When Blériot flew the Channel on 25 July 1909, a flight as long as this – a matter of under 50km (31 miles) – was a marvellous achievement. Throughout the

BELOW *The original Lockheed L-1011-1 TriStar was a medium-haul aircraft tailored to trunk routes in the United States. Today British Airways is one of the operators of the L-1011-500 which has more span, a shorter body, more powerful engines and extremely long range with excellent fuel economy.*

British airways
TriStar

1930s constructors sought to design an airliner capable of crossing the North Atlantic non-stop, but it could only be done – with extreme skill and difficulty – by large flying boats loaded to the brim with fuel and virtually no payload. In the case of the British flying boats they cheated by refuelling in flight to an even greater 'overload weight' after take-off. During World War 2 Britain planned giant aircraft to fly the Atlantic and these eventually flew in 1952 as the Brabazon landplane and Princess flying boat. But by that time the advancing technology of aeroplane structures and propulsion had made them obsolete, and they were never bought by an airline. By 1956 the much smaller DC-7C was flying the Atlantic non-stop, followed a year later by the L-1649 Starliner (ultimate version of the famed Super Constellation and the pinnacle of the piston-engined era) and the rather more advanced Britannia turboprop.

However, in 1958, the jet revolution really started. The jet had first come to the air routes in May 1952 when a de Havilland Comet of BOAC began scheduled services between London and Johannesburg, with five intermediate stops. Tragically, the British firm did not detect a weakness in the primary structure – the highly pressurized fuselage had square cut-outs for the windows and certain radio aerials, and repeated pressurization eventually made the thin skin rip open at the corners – and in 1954 these early Comets were grounded. Thicker-skinned Comets with safer, more rounded fuselage cut-outs followed, but it was not until April 1958 that the first Comet 4 made its maiden flight. This was 50 per cent heavier than the Comet I, carried 81 instead of 36 passengers, and could fly about twice as far with a full load as the 2,500 km (1,550 miles) of its predecessor. On 4 October 1958 the Comet 4 opened the first jet transatlantic service, again with BOAC (which later was merged with British European Airways to form British Airways).

Even the Comet 4 was not really a transatlantic aircraft. It had to make a refuelling stop at Gander, Newfoundland, and ought really to have been con-

LEFT *First of the big jetliners to go into service, in October 1958, the Boeing 707 finally came to the end of the road – as far as new-build civil aircraft are concerned – in 1980, with 962 sold. This 707 is a fan-engined -120B, one of 40 used by TWA on domestic routes in the United States alongside 43 bigger 707s.*

fined to routes to Asia, Australia, Africa and South America where there would have been no range problem. The reason BOAC forced the Comet 4 on the London to New York route was because its arch rival, PanAm, had bought the Boeing 707. This was the first of the so-called American 'Big Jets' – though they do not look particularly big to us today. When they hit the airline scene in the mid-1950s they caused its biggest-ever shakeup – greater even than that triggered off by Britain's Comet. The Comet I paid the price for being first, not only in running into technical trouble (and the answers to that problem, known as structural fatigue, were made available to all its rivals) but also in being almost unwelcome to potential customers.

The airline industry is of necessity conservative and cautious. Though it makes vast sums of money its income is usually almost the same as its expenditure, so the profit is small or non-existent. There is a great incentive to stick to things that are known and trusted. What the airlines like to do is to watch how the new equipment performs with other operators, not for a few months but for years. Then, if the bold pioneering airline can prove satisfaction and profitability, and if it is winning traffic from its competitors, then – and only then – the rest of the world's airlines will join in a stampede for the new airliner.

At first there was no stampede for the jet-propelled Comet. Most airlines found reasons for thinking it radical, unproven, small or unprofitable, and went on buying piston-engined machines. But the worrying factor about the Comet was that jets could fly roughly twice as high and twice as fast. Twice as high meant clear deep-blue sky and utter tranquillity, in place of lurching turbulence and thundering vibration. Twice as fast meant half the journey time, and nobody objected to that. This fantastic passenger appeal made traffic switch to the Comet in droves wherever it flew, and so airlines were slowly forced to join the queue of customers. Had it not been for its fatal flaw it would have sold in hundreds.

However, the Comet emerged into a world dominated by the United States. Such companies as Boeing and Douglas were inwardly deeply worried at the prospect of Britain suddenly taking their place as suppliers of the world's airliners. When another British company, Vickers, won a contract for prototypes of a much larger, faster and longer-ranged jet transport – called the V.1000 for the RAF and VC7 for BOAC – they knew they had to produce an answer. From 1949 onwards they began studying jet airliners to beat the British machines. And on 20 May 1952 the board of the Boeing Airplane Company, as it was then called, voted to commit more than $20 million, virtually the net worth of the company, to constructing a prototype called the Model 367-80. This flew on 15 July 1954, by which time Boeing was talking to the US Air Force about a tanker/transport version to refuel the giant bombers of Strategic Air Command, with the prospect of orders so great as to remove the frightening financial risk that initially had been such a worry. Eventually Boeing delivered no fewer than 732 of the KC-135 tanker/transport to the US Air Force, so that sales of a version for the airlines were almost jam on the bread.

Boeing called its civil variant the 707. Compared with the Comet, even the Comet 4, it was considerably bigger, faster, longer ranged (but still not really a North Atlantic machine) and carried getting on for twice as many passengers. But there was no way it could show any significant advantage over the big Vickers transport. It looked as if the Comet 4 and VC7 would gain at least half the world market by starting off with unrivalled jet expertise, and give the airlines the underpinning of assurance they sought. Then, surprisingly, both the RAF and BOAC decided not to order the Vickers aircraft and this project collapsed in November 1955.

What made this incredible decision all the stranger is that just a month earlier, on 13 October 1955, PanAm had placed an order for 20 Boeing 707s – and, to the great surprise of Boeing, for 25 of a Douglas rival called the DC-8. These now had no competition, and the world's airlines queued up to buy them. On 24 October 1956 even BOAC joined this queue, signing for 15 Boeing 707s!

The first civil 707 flew at Seattle on 20 December 1957. It looked more modern than the Comet, with a much bigger 'double-bubble' fuselage seating passengers in triple seats on each side of the aisle and with what seemed a vast underfloor volume for cargo and baggage. The wings were swept back at 35° and from them hung the four Pratt & Whitney JT3C turbojets in neat pods, with reversers and noise-reducing nozzles that owed much to work by Rolls-Royce. PanAm put in about 130 seats, and began operations with a flight from New York to Paris and Rome on 26 October 1958. Much larger than anything else seen at the airports, and totally different in appearance, the 707 was undeniably impressive, though it was not yet perfect. It was not much more of a North Atlantic aircraft than the rival Comet 4 (puny by comparison), and west-bound flights had to stop at such bleak places as Keflavik, Goose Bay or Gander. A second, and very important point, was that the 707 (and soon the DC-8) did not fit the world's airports. Runways had to be greatly lengthened and strengthened, at a cost of billions, and terminals improved to cope with passengers in much larger numbers. Handling the early 707 was no simple task, and it demanded accurate 'flying by the book' to achieve good results and, for example, to pull up safely on wet runways. As for the social consequences, there was as yet no strong environmental lobby, but the noise and sooty smoke from the early 707s were unforgettable!

ABOVE *Developed in the mid-1950s to rival the 707, the DC-8 was not quite as successful, though the production run was extended by the improved Super Sixty family in the 1960s. This long-range DC-8-63 – an airliner with a long and noisy take-off – is, with a smaller DC-8-50, operated by the national airline of Surinam.*

Douglas's rival DC-8 was even more of a gamble because there was no chance of underpinning it by military orders. The first DC-8 flew on 30 May 1958, and from the start Douglas scored by offering two versions, one with the JT3C engine and the other with the more powerful JT4A unit and considerably greater fuel capacity for longer ranges. In general, a healthy airliner programme is one in which the aircraft offered can be made to develop to greater fuel capacities, greater payload and/or greater range or speed. Whereas the original DC-8-10 had been offered at a maximum weight of 95,700 kg (211,000 lb) and the 'intercontinental' DC-8-30 at 116,600 kg (257,000 lb), the Dash-30 actually grew to 142,900 kg (315,000 lb), which meant greater payload carried over longer distances. Boeing fought back with an enlarged 707 called the 707-320 at 137,000 kg (302,000 lb), which was then certificated at 143,350 kg (316,000 lb) as soon as Pratt & Whitney pulled out more engine power.

In the early days of both aircraft a few were sold with the engine that had been intended for the VC7, the Rolls-Royce Conway. This was the first of a new species called bypass turbojets, with greater fuel economy and slightly less noise. They gave the 707 and DC-8 a little extra range and payload, and they also did no harm to the speed; in fact a Conway-DC-8 exceeded the speed of sound in a shallow dive, the only airliner apart from the Concorde and Tu-144 to do so. But Pratt & Whitney had the inside edge by being American, and by 1960 it also had beaten Rolls-Royce at its own game. By turning the front of the JT3C engine into a large-diameter fan it produced the first of the turbofans, the JT3D. This transformed the 707 and DC-8. Allied with aerodynamic improvements it turned them from marginal vehicles that left three kilometres of smoke behind on take-off, into really sprightly aircraft that could use smaller airports, climb more steeply and stop smoking. With the JT3D, Boeing have continued selling 707s to this day, though the production line at Seattle has almost closed. In the mid-1960s Douglas, part of McDonnell Douglas from 1966, breathed new life

into the fading DC-8 with the Super Sixty series which offered much more capacity (ie, 259 instead of 189 passengers), or greater range, or both.

Eventually the 556th and last DC-8 was delivered in 1972, while Boeing's civil 707 total, including a shorter-range version called the 720, has almost reached 1,000. However, the basic DC-8 is an outstanding aircraft, and many of the later Super Sixty series are to be rebuilt during the 1980s and fitted with a new-technology engine. As outlined in the chapter on how airliners work, high bypass ratio turbofans have revitalized transport aircraft and made them dramatically quieter and more fuel-efficient. One of the newest engines, in the ten-tonne thrust class, is the CFM 56, developed jointly by companies in the United States and France. This engine is flying in a single 707, has been adopted by several airlines – especially all-cargo operators – for the DC-8, and is also to replace the early turbojets fitted to the KC-135 tanker/transports of the USAF that led to the civil 707.

In the 1950s, BOAC, having said it did not want a big jetliner and then buying a fleet of American 707s, soon asked de Havilland to make a rival jet, the DH.118, and then changed its mind and ordered the VC10 from Vickers. It instructed Vickers to tailor the VC10 to the existing short runways around its routes, especially the most difficult ones in hot countries or at high altitudes such as Nairobi and Johannesburg. Thus Vickers had to make the VC10 extremely powerful and give it a large wing and, to enable slats and flaps to generate high lift from tip to tip, the engines were all grouped at the back. BOAC also bought a stretched version, the Super VC10, with even more power and greater range and payload. It ordered ten Supers, changed its mind and ordered 30, then when all 30 were being built cut the order to 17. Once this splendid aircraft was in service, and winning friends among passengers and flight crews for its outstanding grace and quietness, BOAC publicly criticized it and demanded extra subsidies for having to operate it although examination of the accounts failed to support this.

It was the Super VC10 which the Soviet Ilyushin design bureau chose as a pattern in planning the first long-haul Russian jet, the Il-62. Remarkably similar to the British aircraft, the Il-62 first flew in January 1963 and entered service in September 1967. It has had a lot of problems, and more than its fair share of accidents, but it is basically sound and is used not only by the gigantic Soviet airline Aeroflot but also by Eastern bloc countries. Ilyushin also produces an excellent large freighter powered by the same engines as the later Il-62M which plays a big role in civil and military transport for the Soviet Union. The third big Ilyushin, the wide-body Il-86, is discussed in the next chapter because it is, by modern standards, not quite a long-hauler.

BELOW *Outstandingly well engineered and popular with passengers and crews, the BAC Super VC10 served British Airways well for 15 years. By the time this book is published the last 14 will probably have been sold to the Dominican Republic.*

LEFT *Closely similar to the Super VC10, but larger and even more powerful, the Il-62 was the first long-haul jetliner to be designed in the Soviet Union. A little over 130 are estimated to have been built, and no long-range replacement is in sight. This Aeroflot Il-62M was photographed at London Heathrow.*

BELOW *The Boeing 747 is perhaps the best-known of modern airliners. Dramatically bigger, more efficient and quieter than previous jetliners, it caused problems in the early 1970s because its 350-500 seat capacity was too much, both for the existing airports and the traffic. Today Boeing is finding a demand for stretched versions! This colourful Irish 747-100 is seen on take-off.*

INSET *The 747SP is an ultra-long-range version initially sponsored, like the 747 itself, by PanAm, but now sold to 11 other customers. Note its short body and big tail.*

By far the biggest and most advantageous development since the introduction of the jet itself was the advent of the large high bypass ratio fan engine in the mid-1960s. Almost the only company capable of making immediate use of it was Boeing, which produced the monster 747. At one stroke this, the largest, heaviest and most powerful aircraft in service, doubled the payload, increased speed and range and dramatically reduced noise. Even on take-off, with four giant engines putting out 100 tonnes of thrust, there is only a smooth hum – none of the deafening blast of earlier turbojet engines – and not a trace of smoke. PanAm started service with the 747 on 21 January 1970, and though early aircraft encountered a fair amount of bother, and for reasons of size and cost worried the airlines as much as the 707 had done 15 years earlier, the traffic soon grew to match the 747 and we were into the era of mass travel.

Though the various normal 747 versions are among the longest-ranged of airliners, a few oper-

Aer Lingus
PAN AM
PAN AM

British Caledonian
G-BFGI
British Caledonian Airways
AVE 1327 BR
British Caledonian Airways
27

airlines needed extra long sectors and to meet their requirements Boeing introduced the 747SP (Special Performance) in 1973. This has a body 14.2m (47ft) shorter, but a very tall tail, and indication of its capability was provided by delivery of an SP to South African Airways when the flight was made non-stop from Seattle to Cape Town, a distance of 16,560km (10,290 miles), with a heavy load of people and spare parts, and fuel for another 2½ hours remaining at the destination. By late 1980 Boeing had sold more than 550 of all 747 versions, which at over $60 million each is business worth having. Nearly all the early versions had Pratt & Whitney JT9D or General Electric CF6-50 engines, but Rolls-Royce's RB.211-524 is taking an increasing share of the 747 orders because of its superior fuel economy.

These three engines are also used in the two 'wide-body trijets' which were tailored to the requirements of US domestic operators (airlines serving cities within the United States). These enormously busy airlines, led by United, TWA, American and Eastern, at first considered a big twin but later changed to a trijet (three-engined jet), able to use small airfields such as New York La Guardia and fly large loads of passengers to destinations such as Chicago and Kansas City. In 1968 the two trijets were launched. TWA and Eastern bought the Lockheed L-1011 TriStar, powered by the then-new Rolls-Royce RB.211. United and American chose the McDonnell Douglas DC-10, powered by the General Electric CF6. Like all good aircraft these grew in weight and power to meet greater demands for payload and range, and soon both were long-haulers. The DC-10 jumped up into this global league first, because Lockheed was badly hit by the collapse of Rolls-Royce in 1971 and only just managed to survive. But General Electric offered a more powerful engine, the CF6-50, and with it the DC-10-30 was launched with range increased from 4,355 km (2,706 miles) to 7,413 km (4,606 miles). The extra weight, up from 206,400 kg (455,000 lb) to 259,450 kg (572,000 lb), necessitated strengthening the structure and adding an extra main landing gear under the fuselage.

Despite a succession of highly publicized accidents the DC-10 has never ceased to attract orders, and in 1980 new extended-range and increased capacity versions were expected to swell the total further. Lockheed, fighting in the No 2 position, at last managed to launch various later variants of the technically superb TriStar, the most important being the long-haul L-1011-500 which has much more powerful RB.211 engines and is generally regarded as the most fuel-efficient long-hauler on the market. Lockheed have pioneered the use of 'active controls' which automatically deflect the ailerons on the wingtips to oppose the up-thrust of a gust, a suddenly encountered column of swiftly rising air (in the old days gusts were called bumps). Flying into a strong gust at 800 km/h (500 mph) puts a great strain on the wings, which are sharply bent upwards, but active ailerons greatly reduce this effect by bending the tips downwards. Thus the wing can be extended to a greater span without overstressing the structure, and a longer, more slender wing means greater efficiency and more range on the same fuel capacity. Active controls will soon be found on the majority of the world's advanced airliners.

LEFT *First sold, at the same time as the rival TriStar, in 1968 (to American and United) the McDonnell Douglas DC-10 then drew ahead to become the best-selling wide-body after the 747. It gained from Lockheed and Rolls-Royce troubles in 1971, and then scooped the pool of long-haul orders at a time when General Electric, but not Rolls-Royce, could offer a more powerful engine, matched to higher gross weights and greater fuel capacity. This DC-10-30 of BCal is one of these long-range versions. In recent years various factors have caused sales to fall off, and the maker has probably missed the chance of launching a stretched DC-10-60 seating as many as a 747.*

Alitalia
A300

SHORT/MEDIUM-HAUL JETS

In the early 1950s, when airlines were just beginning to take the pioneer de Havilland Comet seriously, it was generally agreed that even for long hauls – which in those days meant anything much over 2,000km (1,250 miles) – the turboprop had to be superior. Jets, it was argued, could show a substantial speed advantage, but this was largely negated by the cost to the operator of the extra fuel on each trip.

When it came to short hauls, the jet was barely discussed at all. On sectors of about 2,000km or less it could hardly show a significant saving in journey time. The cruising part of the flight might save five to ten minutes, but the rest of the time – the taxiing round the airport, take-off and climb, descent and taxiing at destination – appeared on close examination even to favour the turboprop, which could fly at full speed low down and did not have to climb so high. Consequently, there seemed no point in considering the matter any further. It seemed pointless to buy a costly, noisy, smoky, fuel-hungry jet for short flights when there was no evident benefit.

BELOW *When Boeing at last went ahead with the 767 in July 1970 it faced a totally new situation: large areas of the world had already bought the rival Airbus products. One of the few regions still to be fought for was the Middle East, and Airbus has since won most of this too. Clearly, the European aircraft is hard to rival. This A300B is a GE-engined aircraft of the national airline of Italy.*

One of the very first jetliners offered to the airlines had been a short-hauler. Avro Canada at Toronto had an outstanding engineering team which in 1946 began work on the C.102 Jetliner. A thoroughly modern design seating up to 52 passengers, it had 'straight' (unswept) wings, powered flight controls and a roomy pressurized fuselage. Technically its only shortcoming was the four rather outdated Rolls-Royce Derwent engines in twin underwing nacelles. The Jetliner flew just two weeks after Britain's Comet, on 10 August 1949. Although it was a complete success no orders followed and the Jetliner just faded away.

The Canadians had paid the penalty not of being wrong but of being too early, which in the airliner market is as bad as being too late. Yet in the early 1950s several design teams studied short-haul jets. In East Germany a somewhat primitive aircraft was actually built and flown, notable for its similarity to the American B-47 jet bomber with tandem main landing gears under the fuselage, a high wing and twin-jet pods hung below it on pylons. The first short-haul jet in the United States was even cruder: two twin-jet pods were hung under the wings of what had been a piston-engined freighter! But in France a new jetliner was being planned with all the skill the Gallic designers could muster, and this was destined to break through the almost closed minds of the customer airlines. More than that, it triggered off the supposed supremacy of the jet even on short hauls, so that airlines quickly came to believe that anything with propellers was old fashioned.

At first the French project was called the X-210, but in January 1953 the designation was changed to SE.210, to reflect the name of the builder, SNCASE, a quite small company with little experience of airliners. Other French groups were developing a turbojet, the Atar, and it was planned to use three of these in a group at the tail. Perhaps wisely this idea was dropped in favour of just two of the more powerful British Rolls-Royce Avon engines, but the novel rear-fuselage location was retained. Another new idea was a stairway under the tail leading up to a door at the back of the interior. There was no argument about the fuselage cross-section; to save time the complete nose and flight deck of the British Comet was adopted. Other features included highly efficient long-span wings with airbrakes and spoilers, bogie main landing gear (remarkably short ones that placed the aircraft quite close to the ground), fully powered flight controls with the horizontal tail a short way up the fin, and unique triangular windows that gave passengers a good view downwards and yet whose generous rounded corners avoided the fatigue weakness that later caused problems on the Comet.

By the time the first prototype flew, on 25 May 1955, the aircraft had been named Caravelle. With solid backing by Air France, European operators such as Swissair, SAS of Scandinavia and Sabena of Belgium, everyone stopped thinking of short-haul jets as silly and thought instead of passengers transferring their allegiance to the new airliners that could offer such swift and effortless travel. The Caravelle became the most important French civil aircraft programme, and eventually in 1972 the 282nd and last example of this highly successful airliner was flown away to its customer.

It established an excellent record in many versions which progressively offered greater range and payload. Even the first aircraft had seated more passengers than the early Comets, 64 being a typical number, and the final model could carry as many as 140. The only other aircraft to pioneer short-haul jet travel was an amazingly simple conversion of a bomber by the Tupolev design bureau in the Soviet Union. While many constructors had studied airline derivatives of jet bombers, only the Tupolev bureau was really in a position to do it, because its Tu-88 bomber was ideally arranged for such a conversion. To produce an airliner all the Russian designers had to do was build a much larger circular-section pressurized fuselage and mount this completely above the wing, so that the centre-section spars did not obstruct the cabins. This made the nose landing gear longer, and that was almost the only other change.

The first of these remarkable civil conversions, designated Tu-104, flew on 17 June 1955. In April

RIGHT *The first true short-haul jet to go on the drawing board (except for the premature Avro Canada Jetliner of 1949), the French Caravelle was also the pioneer of rear-mounted engines which improved wing efficiency and reduced noise in the cabin. This Spanish Caravelle was one of 12 Series VI-R examples with British Avon engines bought by Iberia. The other two aircraft in the picture are early BAC One-Elevens of newly merged British United and Caledonian.*

1956 one flew to London, causing an immense stir among Western aviation people who had never seen a large Russian jet at close quarters before. On the whole they found it very workmanlike, though the glazed 'bomber-type' nose and under-nose radar excited comment. Even today most of Aeroflot's airliners have glazed noses because they still carry old-fashioned navigators who like to see the ground ahead and map read to their destination in the vast regions of the Soviet Union where there are few electronic navaids. At first the Tu 104 was uneconomic, seating 48 passengers and burning a lot of fuel. Later versions, of which as many as 200 were built, could accommodate up to 100 and for 20 years provided high-speed transport on many routes from Paris to Vladivostok. Tupolev also produced a scaled-down model, the Tu-124, which was much more economical though it also carried fewer passengers, the usual number being 56. Again, about 200 were built, though all had been retired by 1981.

In the mid-1950s a big American company, General Dynamics, decided to take a giant gamble and build a short-haul jet. It was already important as a builder of piston-engined airliners with the Convair 240, 340 and 440, and it decided to build a jet similar in layout to Boeing's 707 but rather smaller and with much less fuel capacity. This eventually matured as the Convair 880, and the first off the production line flew on 27 January 1959. But even including the 880M extended range version total sales amounted to a mere 65, largely because Boeing was able to undercut the price with the 720 and fan-engined 720B which were lightened short-range versions of the 707. General Dynamics could have given up, but instead built a larger and faster model, the Convair 990 Coronado. Advertised as the fastest airliner in the world, this flew on 24 January 1961, but failed to achieve the guaranteed performance. It took two years of extremely expensive modifications, in the course of which four large 'shock bodies' were added above

RIGHT *Boeing's vast customer list of just over 100 names for the 727 includes giants such as the major US domestic trunk airlines, several of which have fleets of over 100, and small operators with a single aircraft. The biggest fleet is that of United, the world's largest airline except for Aeroflot. Shown here is one of United's 186 727s.*

the trailing edge of the wing to improve airflow, before fully modified 990A airliners entered service in late 1962. The corporation lost well over $450 million building these two jetliners, one of the greatest ever commercial disasters!

Meanwhile in Britain de Havilland merged with the giant Hawker Siddeley Group and received an order for its DH.121, an extremely modern trijet to meet BEA's needs. Surprisingly the airline then changed its mind and made the company cut down the size, which in turn forced Rolls-Royce to redesign the engine to give only 4,468 kg (9,850 lb) thrust instead of 6,350 kg (14,000 lb). The resulting engine was a neat bypass jet called the Spey (at that time Rolls-Royce named its engines after rivers); the restyled 121 was renamed Hawker Siddeley Trident. Extremely fast, and abounding in advanced features such as four small mainwheels on each axle and a T-tail with the horizontal tail atop the squat fin, it seated 88 and had a unique advantage in a Smiths Autoland system enabling it, in stages of development that took place during the 1960s, to land safely in completely blind conditions. The first Trident flew on 9 January 1962, and service with BEA began more than two years later on 11 March 1964.

The next version of Trident, the 2E, was developed for longer sectors such as London to Athens non-stop, and the ultimate version of the Trident was stretched to seat up to 180 passengers, 70 more than the original DH.121 which had frightened BEA because they thought it far too large! Though the small Spey had been developed to give 5,425 kg (11,960 lb) thrust, this was still not enough for the enlarged Trident 3 and eventually the novel solution was chosen of adding a take-off booster jet. The RB.162 is a simple turbojet originally designed for VTOL (vertical take-off and landing) aircraft, but in the Trident 3 it is mounted in the normal fashion in the tail, above the main centre engine and fed from inlets in the side skins which are normally covered by doors. Each RB.162 operates only during take-off and the climb away from the airport.

LEFT *Originally designed exactly to the requirements of BEA (now part of British Airways) the Hawker Siddeley Trident was later developed in larger and longer-ranged versions, but failed to offer competition to the 727. Largest model is the Trident 3, in which the need for extra thrust at take-off was cunningly (if noisily) met by adding a fourth engine. The nozzle for the small RB.162 booster jet, used only on take-off, can be seen above that of the centre engine.*

Altogether 115 Tridents were built, but this compares with more than 1,900 of its direct rival, the Boeing 727. Started much later than the DH.121 and in service ahead of the Trident, the 727 was almost exactly like the original de Havilland project. The only differences were subtle, notably that the 727 was not tailored to just one airline but built for the world market. One important way in which it differed from its British rival was that the wing was made more advanced, with exceptional high-lift slats and triple-slotted flaps, so that even though the 727 was considerably heavier that the Trident 1 it could use smaller airports. When the first 727 flew on 9 February 1963 it demonstrated a lower weight of structure and higher flight performance than had been calculated, and while taking orders hand over fist Boeing began a product-improvement programme which introduced the 727-200 stretched to seat 189 passengers (the original 131-seat model becoming the -100), the 100C all-cargo model and 100QC (Quick Change) which normally carries passengers in luxury by day and quickly changes to cargo by night, and finally an Advanced 727-200 with even greater fuel capacity and many detail improvements.

At first the 727 suffered a succession of accidents, most of them caused by pilots allowing the

ABOVE *A fine portrait of the most successful British jetliner, the BAe (originally BAC) One-Eleven. This Series 500 equipped with noise-suppressed engines was one of the last to be assembled in Britain, at Hurn, near Bournemouth, where No 245 was completed in 1981. No 246 and as many as 100 further aircraft are expected to be assembled in Romania, with increasing local content until by 1986 the airframes will be entirely Romanian.*

speed to fall too low on the landing approach so that they got into a steep and dangerously rapid descent which not even maximum engine power could arrest. Once they flew strictly 'by the book' these accidents ceased, and from 1966 onwards the only criticisms that could be levelled at the 727 were that it was noisy and poured black smoke. Pratt & Whitney, makers of the three JT8D turbofan engines, continued not only to produce more powerful versions but also to offer airlines a cleaned-up combustor which almost eliminates visible smoke, and this is fitted to all later 727s. By modern standards the 727 is obsolescent, and Air France found it uses 30 to 40 per cent more fuel per passenger on its routes than the Airbus A300B. Depite this the airline industry is so totally hooked on vehicles that it knows and can rely upon, that the Boeing 727 is still selling briskly for delivery as late as 1983!

After the Caravelle the first modern twin-jet was the British BAC (now British Aerospace) One-Eleven. In fact it was a project of Hunting Aircraft at Luton, one of the companies taken over by BAC when it was formed in 1960, and it was tailored around two Spey engines to meet the needs of Freddie (now Sir Frederick) Laker, then managing director of British United (now called British Caledonian). By the time the extremely attractive prototype flew on 20 August 1963 the design had grown to seat 79 passengers, and it was then exhaustively tested in the course of which it sadly suffered a 'deep stall' crash. However, BAC's eagerness to discuss this openly and tell rival builders its findings certainly helped others avoid such a catastrophe. The One-Eleven was exactly right, and the developed Series 400 with extra fuel and American equipment was even adopted as standard short-haul jet by American Airlines. Other US customers included Allegheny and Mohawk, whose short-body Series 200 aircraft now fly under the big grouping US Air. Most One-Elevens now in production are part of a long-term deal with the aircraft industry of Romania which is expecting to make advanced versions throughout the 1980s.

Successful penetration of the US market by the One-Eleven triggered a response by Douglas Aircraft. This famous airliner builder had offered a DC-9 in 1958, looking like a much smaller DC-8, but never cut metal. When the DC-9 finally went ahead in April 1963 it was a totally different aircraft almost exactly like a One-Eleven with rear engines and a T-tail, but slightly larger because of the greater power of the JT8D engine. The first flight took place on 25 February 1965, and after an impressively fast development programme Delta Airlines put the DC-9 into service on 8 December the same year. Since then McDonnell Douglas (which absorbed Douglas in 1966) has sold just 1,000 of many versions which extend from the 72-passenger DC-9-10, with an overall length of

BELOW *In 1980 the baby of the Boeing range, the stubby 737, not only overtook the 727 in sales but also unexpectedly appeared in brochures offering a 737-300 version with completely new fuel-efficient engines. There seems every chance this 'baby Airbus' may by 1986 reach the 2,000 mark.*

AUSTRIAN

BELOW *McDonnell Douglas have stretched the DC-9 more than any previous airliner. In five stages passenger capacity has been increased from 72 to 172, the biggest being this impressive Super 80 version which went into service in 1980. Not least of the Super 80's improvements is the engine which, though derived from the original JT8D used in earlier versions, has been redesigned to give more power, be more fuel-efficient and make less noise near airports.*

31.82m (104ft 4¾ in) to the largely re-engineered DC-9 Super 80 with a length of 45.06m (147ft 10in) and seats for 172 passengers! No other airliner has ever been made in such a wide range of sizes.

On 19 February 1965, when the DC-9 seemed to have sewn up the US short-haul jet market, Boeing suddenly announced it was going ahead on the 737, for which the first order had been most unusually placed by a foreign airline, Lufthansa. Using many parts from the 707 and 727, and thus having a relatively portly body, the 737 had its two JT8D engines tucked under the inner wings immediately outboard of the landing gears. The 737-100 seated 100, looked extremely stubby, but from the start of service on 10 February 1968 (just ten months after first flight) it proved reliable and popular. Predictably the next move was a stretched 737-200, seating 130, and today this is the fastest-selling of all the small twin-jets with the total approaching 900 despite the fact the engines are of an obsolete generation. (Pratt & Whitney has developed a more modern 'refanned' JT8D, giving greater power and better fuel economy while considerably reducing noise, and this has been selected for the DC-9 Super 80 and was in 1980 being studied by Boeing for a proposed 737-300 family.) Among the many features of the 737 are front and rear doors with built-in airstairs, a range of quick change and executive versions, and an optional kit for allowing operation from gravel and other unpaved surfaces which among other items includes a deflector behind the nosewheels and high-pressure air jets ahead of the main engines to prevent ingestion of stones and other hard objects.

In the Soviet Union Tupolev followed the 104 and 124 with the rear-engined, T-tailed Tu-134 and the much larger 154, even bigger and heavier than the Advanced 727-200 and with six-wheel main gears to spread the load on soft airstrips. To meet Aeroflot's need for a large-capacity passenger transport, the Ilyushin bureau in 1972 planned the Il-86 with a wide circular-section fuselage, four rear engines and a T-tail, but when the prototype flew in December 1976 the configuration had been

changed to that of the 747. Three stairways lead from the ground to the underfloor area where passengers stow baggage and outdoor clothing. Three more stairways then lead to the main deck, typically seating 350. Another Russian aircraft is the Antonov An-72, whose two turbofans blow across the wing and flaps to give STOL (short take-off and landing) performance, enabling it to operate from short airstrips.

In Western Europe Fokker produced the F28 Fellowship, again a rear-engined T-tailed twin but one planned with care to meet the needs of smaller operators with smaller airports. Noted for its large rear-fuselage airbrakes, the F28 has been developed in many versions seating 65 to 85 and is selling in ones and twos to numerous customers. But two other West European short-haulers lost their manufacturers vast sums: France's Dassault-Breguet Mercure, similar to the 737, was an international collaborative project for which several new factories were built but only ten aircraft were sold to the French airline Air Inter, while West Germany's baby VFW 614, with Rolls-Royce M45 turbofans mounted on pylon struts above the low wing, was abandoned at the end of 1977 after only 16 had been built.

In total contrast, Europe's biggest-ever civil aircraft programme, and the only one in the world to challenge US domination, has been a predictable smash hit. In the late 1960s it was blindingly obvious that, if high bypass ratio turbofans and wide bodies made sense for long hauls, they also did for shorter routes where most of the traffic travels. Airbus Industrie was formed in 1970 to combine the efforts of France, West Germany and the Netherlands in building a big twin to carry about 250 people on typical European or US sectors. Britain regrettably pulled out but later Spain joined, in 1978 British Aerospace became involved (previously Hawker Siddeley had designed and built the wings at its own risk) and Belgium joined to work on the latest AI product, the A310.

The original design, the A300B, is the quietest and most fuel-efficient jet in the world, with ad-

ABOVE *The Tupolev Tu-134 first flew in 1962 and the stretched Tu-134A was still in low-rate production at the Kharkov plant in 1980 to make up for the delay in getting the Yak-42 into service. Typical Russian features suiting it to rough environments are the eight big low-pressure tyres for poor runways, retracting into wing pods, and the glazed nose for visual navigation. This 134A is one of a fleet of 11 (four of them short-body 134s) of the national airline of Poland.*

vanced-technology engines and a supercritical wing. It first flew on 28 October 1972 and entered service with Air France on 30 May 1974. But, as it was not American, almost all the customers stayed away. Sales were gained with great effort in ones and twos to a few far-sighted airlines such as those of France, West Germany, Korea and South Africa. Then in 1979 the airline industry suddenly realized the A300B was not a mirage and that its multi-national producer was not going to vanish overnight. Sales shot up to over 380, and have since climbed to beyond 420, though the problem now is that the AI partners could find it difficult to meet demand. The A310 is a shorter model in the 200-seat class with a new small wing. The consortium is also considering stretched 350-seat versions and 132/180-seat SA (single-aisle) 'narrow bodies'.

Boeing studied new short-haul jets for ten years before launching the narrow-body 757, seating up to 233, and the totally different 767 with a body narrower than an Airbus but still called 'wide', seating up to 255. The 767 has been planned for airlines suffering from short runways or other adverse circumstances because – at a penalty in operating cost – it has a wing not only larger than that of the A310 but even bigger than that of the far more capable A300. Another odd feature is that the smaller body cannot carry standard underfloor cargo-baggage containers, yet Boeing's reputation is so great that airlines are undeterred. For example, virtually the only national airline in non-communist Europe not to have bought the Airbus is that of Britain, an AI partner; yet British Airways has bought the American 757!

RIGHT *The most up-to-date flight deck at the time of publication is that of the Boeing 757 twin-jet, a very long narrow-body with modern engines. Instead of a mass of dials the two pilots will have large electronic displays on which they can call up just the information they need. Originally this aircraft was intended to have a flight deck based on that of the old 727, but now it is closely related to that of the wide-body 767.*

LEFT *One of the world's best freighters is the Russian Il-76, built in large numbers for military and civil use. Despite the fitting of comprehensive electronics the nose remains glazed, both for visual navigation and to assist accurate air drops through the rear doors (seen open in this photograph). Very powerful engines and multi-wheel landing gear are among features facilitating operation from short strips, for example in opening up Siberia.*

ABOVE *Yakovlev's Yak-42 was the first Russian aircraft known to have modern high-bypass-ratio fan engines. Rather like a Trident but tailored to rough airfields, it first flew as long ago as March 1975 but was still not in service in early 1981.*

There is today a stern battle developing between the advanced, quiet, fuel-efficient turbofans and the turboprops discussed in the next chapter. In 1968 the Soviet Yakovlev bureau flew the first Yak-40 tailored to really small 'outback' airstrips; yet instead of turboprops it fitted three rear-mounted turbofans to this straight-winged, slow-flying 33-seater. Yakovlev later built the faster, swept-wing Yak-42, again a T-tailed trijet, seating 120. Fokker's next major project, the F29, is a jet in a slightly larger size class, and McDonnell Douglas has been trying to launch an ATMR (Advanced Technology Medium Range) seating about 188 and supposedly more efficient than the 757. Perhaps the greatest success will be won by an aircraft which has begun life like the Airbus surrounded by doubters and critics. The British Aerospace 146 is intended as a natural replacement for old turboprops and is offered seating up to 88 (146-100) or up to 109 (146-200). Powered by four fuel-efficient Avco Lycoming ALF 502 geared turbofans, it is tailored to the shortest and most difficult airfields and yet has outstanding all-round qualities including near zero noise pollution. Risk-sharing partners Avco Aerostructures (USA) make the wing and Saab-Scania (Sweden) the tailplane and control surfaces. Orders from Argentina and the USA in mid-1980 are certain to be the first of many.

TURBOPROPS

BELOW *Really good pictures of Eastern Bloc aircraft are rare, so this scene of passengers disembarking from an Il-18 is welcome. In the class of the Britannia, but not originally intended for long ranges, the Ilyushin passenger liner has been made in large numbers since 1957. A feature prominent in this view is the sharp line of discontinuity between the pressurized fuselage and the nose fairing.*

A turboprop is essentially a turbojet arranged to drive a propeller. All the earliest forms of gas-turbine engine were intended to provide power at a rotating shaft, and studies for turboprops go back to the 1920s. It was the great mental leap of RAF pilot Frank (later Sir Frank) Whittle in 1930 that a gas turbine could generate thrust without a propeller that led to the jet age. The most important advance brought by the turbojet was escape from the limitations on speed imposed by propellers, and such engines were first used in high-performance military aircraft. But, alongside the booming technology of jets, engine designers also continued to study turboprops.

Obviously a turboprop is inclined to be heavier and both more complex and more expensive than a turbojet because it consists of the same components as a turbojet (compressor, combustor and turbine) plus additional turbines that extract as much of the available power from the flow of hot gas as possible and use it to drive a propeller. On the other hand for a given size of engine and rate of fuel consumption a turboprop can generate much more thrust at take-off and low speeds than a turbojet, and do so with very much less external noise.

The first practical turboprop was literally an early turbojet (a Rolls-Royce Derwent) to which had been added a gearbox and drive to a small five-bladed propeller. This lash-up engine, called the Trent, took to the air in a Meteor fighter in September 1945. By this time Rolls-Royce was developing a purpose-designed turboprop, the

Dart, and this engine flew in the nose of a Lancaster in 1947. It was largely because of the company's reputation that the Dart was picked by Sir George Edwards of Vickers-Armstrongs to power his new turboprop airliner the Viscount, in preference to rival engines from Armstrong Siddeley and Napier. The Dart was in the 1,000 horsepower class, and the Viscount had four of these tough and mechanically simple units in pencil-slim nacelles driving neat four-blade propellers specially designed by the Rotol company.

ABOVE *Though it took so long to develop that it emerged into the age of the big jetliners, the Bristol Britannia eventually became a mature and capable long-hauler of which a handful survive to this day. This ex-RAF Mk 253 is the sole equipment of an airline based in Ghana.*

The prototype Viscount flew on 16 July 1948. It was in all respects a most modern design, with twin-wheel tricycle landing gears and a pressurized fuselage which had been enlarged from the original 24-seat design to accommodate 32 passengers. It was clearly a design of immense potential, and in the summer of 1950 the prototype was actually put into regular passenger service by British European Airways (BEA, now part of British Airways) between London and Edinburgh and Paris, with exceptionally favourable comments by everyone concerned. At first most airlines had not only fought shy of anything as radical as a turboprop liner but also had been unimpressed by the Viscount's economics. But predictably Rolls-Royce kept pulling more power out of the Dart. At a level of 1,400 horsepower Vickers built the enlarged Viscount 700, and this flew in April 1950 and was ordered by BEA four months later. Seating 44 to 53 passengers, it was much quieter, smoother, faster and more capable than any rival, and an incidental appeal was provided by the enormous elliptical passenger windows. Yet another immense advantage, which was appreciated only over the subsequent decade, was that Dart power made the Viscount amazingly reliable and trouble free, just the opposite of what airlines expected.

Orders trickled in to the Weybridge factory, but in November 1952 Trans-Canada (now Air Canada) bought a fleet of 15 with North American engineering features, and in 1954 Capital Airlines (later absorbed into United) of Washington took the plunge and actually imported this 'foreign airplane' for use on US domestic routes. Eventually Trans-Canada bought 51 and Capital 60, by far the biggest-ever British exports of airliners. In 1956 Vickers flew the stretched 800-series, seating over 70 passengers (three times the original capacity!) and from 1957 nearly all were of the 810-series with 2,100 horsepower engines and cruising speed increased to at least 560 km/h (350mph). By 1964 BAC, into which Vickers had merged, had completed 444 of these fine aircraft, of which 438 were for customers. More than 150 are still in airline use; but even more remarkable is the fact the intrinsically ancient Dart engine – described affectionately by one of its customers in 1959 as 'agricultural machinery' – is still in brisk demand for turboprop airliners all over the world.

Chronologically the next important turboprop liner was the big, long-haul Britannia, planned in 1946 with piston engines and accommodation for 32 to 50 passengers, and eventually flown with four 3,500 horsepower Bristol Proteus turboprops in August 1952, sized to seat as many as 92. In almost all respects it was a superb airliner, especially when the Bristol company stretched it with 4,450 horsepower Proteus engines and integral-tank outer wings for transatlantic range and up to 139 passengers. Cargo versions were developed, but

Bristol took so long (five years) to get the aircraft into service that only 60 were sold, plus 23 military versions for the RAF. A few survive, and these are mainly used for cargo charters such as racehorse transport. Many later versions were planned but the only one to materialize was the Canadair CL-44 built at Montreal with Rolls-Royce Tyne engines. Some of these were of the CL-44D-4 version with a complete hinged tail to allow pallets and outsize freight to be loaded straight in at the rear. Nearly all are still in use, some having cabin windows and provision for up to 214 passenger seats.

A single example, designated CL-44O, was rebuilt in 1969 with an outsize upper lobe to the fuselage to enable it to carry larger items of cargo. This kind of conversion was pioneered by a US company, Aero Spacelines, using as the basis the Boeing Stratocruiser piston-engined passenger liner. Known as the Guppy family, these amazing rebuilds only continued the process followed by Boeing in 1944 in producing the Stratocruiser from the B-29 bomber. At that time the transport fuselage seemed gigantic, but in 1962 the first Guppy looked an absolute monster, designed for the carriage of space rockets and similar outsize loads. This retained its piston engines, but most of the subsequent versions had T34 or Allison 501 turboprops giving much greater power, to fight the drag of a gigantic upper fuselage lobe giving an unobstructed interior 7.7m (25ft 6in) in width or height. Some Guppies have swing-tails but the biggest have hinged noses which pivot round under hydraulic power complete with the nose landing gear and flight deck. The two most advanced models, designated Guppy 201, are used to bring the parts of Airbus A300Bs, and military Transall turboprop freighters, to Toulouse for final assembly.

BELOW *The amazing series of Super Guppy carriers for outsize loads was designed by Aero Spacelines of Santa Barbara, California, using wings, landing gear and some other parts of Boeing 377s or KC-97s. This is one of two used to ferry parts of Airbus airliners between the member companies' factories. UTA at Le Bourget (Paris) is now making two more to help increase Airbus production.*

LEFT *Most successful large airliner of British design, the Viscount brought speed, comfort and reliability to short-haul operation. This V.806, one of the later and more powerful models, flew with a UK airline merged into British Airways.*

In the period immediately after World War 2 the great Dutch planemaker, Fokker, sought to get back into the airliner business by producing a successor to the omnipotent DC-3. In August 1950 it drew up project study P.275 for an all-metal high-wing 32-seater with two Dart turboprops. By 1952 this had matured as the F27, seating 40 in ten rows of four and with double-slotted flaps to match good field performance with high cruising speed and a long-span wing. The first F27, later to be named the Friendship, flew on 24 November 1955. Surprisingly, the overall performance was so good that the flaps were simplified to the single-slotted type, and with a few further improvements the F27

ABOVE *Best-seller of the West's turboprops, the Fokker F27 was designed with a wing optimised not for short airfields but for efficiency in cruising flight. Far from running down in production the rate of output was in 1981 being stepped up to meet demand.*

went into production at Amsterdam and also at Baltimore in the USA as the FH-227, licence-built by Fairchild. Airline service actually began with one of the American-built aircraft, on 27 September 1958, followed by a Fokker F27 with Aer Lingus in December of that year. Since then the F27 has become the best-selling airliner ever developed in Europe, with well over 700 sold (including 128 made by Fairchild) and production increasing to a planned 30 per year. Fokker even expects the F27 to outlast the F28 jet and remain in production to the end of the century.

While Fokker was designing the F27 a British company, Handley Page, was designing a similar machine, the Herald. After talking to many airlines it chose four piston engines, but the unexpected success and sales appeal of the Dart engine forced an agonizing reappraisal. The Herald had flown in August 1955, ahead of the F27, but by the time it had been redesigned to use two Darts it was running in second place, and it never really caught on despite being a thoroughly sound machine. Only 48 were delivered, most now with Air UK.

With the F27 and Herald established in service, and a trim low-wing airliner with twin Darts (the Aviation Traders Accountant) failing to find a market, it was a shock when a third twin-Dart machine, superficially extremely like the Accountant, was announced as the Avro 748 in 1959. Avro was part of the Hawker Siddeley Group and for most of its life the type was called the HS.748; today it is the BAe (British Aerospace) 748. Differing from the F27 in having a low wing, the 748 slowly gained a toehold, and proved especially popular among military and para-military customers; for example it is believed to be used by more heads of state, including the British Royal Flight, than any other aircraft. Gradually the list of customers grew, until by 1980 the 748 was in airline service in a remarkable 48 countries, with sales approaching 400. The new 748-2B has extended wingtips for better performance in hot or high-altitude airfields, and the latest Dart 536-2 engines have noise-suppressing inlets.

Superficially very like the 748, but powered by Darts of the more powerful Mk 542 type rated at 3,060 horsepower instead of the regular 2,100 horsepower, the NAMC YS-11 is Japan's only successful airliner. Seating as many as 60 passengers, compared with a limit of 56 for the F27 or 748, the YS-11 was sold to the tune of 182 examples in several models, including cargo and mixed-traffic versions. Also powered by this powerful Dart engine are the Convair 600 and 640, which are re-engined CV-240 or CV-340/440 piston-engined airliners of the 1948-56 era. The same types re-engined with the American Allison 501 are designated Convair 580, some 130 being produced.

The only really successful turboprop in the high-power class produced in the United States, the Allison 501 was selected by Lockheed in 1954 to power the only American turboprop airliner bigger than small commuter types. In 1954 the Allison 501 was becoming a well-proven engine, and though Lockheed looked carefully at jets these seemed too much of a gamble and too large a financial investment. US airlines, led by American, suddenly found themselves unable to meet the speed and passenger appeal of the British Viscount. Convair, whose 240/340/440 airliners were hit badly, even proposed a four-Dart aircraft called the Convair Dart. However, it was never built, and the way was open for Lockheed to build a really modern turboprop liner, bigger and faster than the Viscount and seating up to 100. The L-188A Electra flew on 6 December 1957. With four 3,750 horsepower Allison 501 engines turning propellers with four extremely broad rectangular blades, it performed in sprightly fashion, often giving demonstrations with three propellers feathered and able to cruise at 652 km/h (405 mph). At first flight 141 aircraft had been ordered, but this was just the time that the airlines suddenly became hooked on short-haul jets. It did not help when two fatal crashes, caused by serious structural failure in cruising flight, led to a major redesign of the engine installation and wing, and only 161 were sold altogether.

About a year later came the Vickers Vanguard, designed to BEA's requirements as a much faster and larger successor to the Viscount. Even bigger and faster than the Electra, the Vanguard could claim to be a second-generation aircraft, carrying 139 passengers at 684 km/h (425 mph) on four 5,545 horsepower Rolls-Royce Tynes. The first Vanguard flew on 20 January 1959, and a substantial order was placed not only by BEA but also by Trans-Canada; but no turboprop was hit so hard by jets as this outstanding machine and only 43 were sold, making it a disaster for Vickers-Armstrongs. Today, when virtually all the surviving Vanguards, and many of the Electras, have been converted into freighters, they are belatedly being recognized as fuel-efficient workhorses.

In the Soviet Union fashion plays little part in major procurement policies, and there the attributes of the turboprop were never in doubt. The first Russian turboprop airliner was the biggest, most powerful, fastest and longest-ranged turboprop in airline history. The engine was the Kuznetsov NK-12, developed mainly by a team of German prisoners in the immediate post-war era and eventually cleared at the remarkable output of 15,000 horsepower, transmitted through a monster contra-rotating propeller with two sets of four blades. Four of these immense engines were used by the Tupolev bureau for the Tu-95 bomber, called Tu-20 by the Soviet Air Force (and dubbed 'Bear' by NATO), first flown in 1954 and still a very important service type in 1980. In 1957 the prototype was completed of the Tu-114 transport version, with a much larger fuselage mounted above the swept wing just as had been done with the transformation of the Tu-88 into the Tu-104. Though the interior decor of mahogany, brass and lace was outdated, the flight performance of this monster was incredible, including lifting 30-tonne loads to over 12 km (39,400 ft), distances exceeding 10,000 km (6,200 miles) with a 10-tonne load and speeds up to 871 km/h (541.5 mph) over a 1,000-km (620-mile) circuit. A non-stop flight was made from Moscow to Irkutsk and back (8,500 km, 5,280 miles) and soon the Tu-114 was in Aeroflot service, with 170 passengers housed in numerous separate compartments, on long routes across Siberia and even as far away as Cuba.

Two other important turboprop liners are the An-10 and Il-18, both in the class of the Britannia or Vanguard with four AI-20 engines. The high-wing Antonov product has been built in large numbers in An-12 freighter form, the An-10 being a rare passenger machine. The Ilyushin aircraft is one of a few Russian types to have been exported widely, over 100 being sold at prices about half those charged by Western builders to numerous smaller operators. Seating up to 122, more than 800 were built in all from 1957. Even larger numbers, certainly exceeding 1,000, were made of the An-24, a high-wing transport in the class of the F27 and Herald but with greater power, though not having better payload or range. A more powerful version is the An-32, tailored to operations in adverse hot or mountainous regions, while the An-28 is an even more STOL (short take-off and landing) 15-seater able to fly anywhere on Aeroflot's enormous route network, with a long-span high wing and two engines in the 1,000 horsepower class. It is one of the few modern aircraft with a twin-finned tail.

BELOW *Like rival Fokker, British Aerospace is working hard on improvements to the 748 to increase the appeal of a fuel-efficient turboprop which is in any case selling strongly. This one works for British Airways Scottish Division.*

SP-LTH

RIGHT *Pictured over the Canadian Parliament Buildings in Ottawa, the de Havilland Aircraft of Canada DHC-6 Twin Otter is the most widely used small turboprop commuter, with nearly 800 sold. Normally seating 20, Twin Otters are found with fixed wheels, skis or floats.*

PRECEDING PAGES *Though Lockheed never succeeded in selling civil versions of the larger, turbofan-engined C-141 and C-5A, civil L-100 models of the reliable and long-established Hercules remain in production, like the C-130 military variants. Because civil loads tend to have lower density, the L-100 models are stretched to give more capacity. This L-100-30 is one of a fleet of 15 (plus a shorter L-100-20) of Safair Freighters of Johannesburg.*

PRECEDING PAGES, INSET *When production ended in 1978 about 1,100 Antonov 24s had been built, a total exceeding that of any other turboprop airliner. This one is Polish.*

CF-CST

RIGHT *Thanks to careful attention to customer demands Short Brothers of Ulster have a best-seller in the 30-seat Model 330. Comfortable and quiet, yet priced at a competitive £1.3 million, orders were fast approaching 100 by 1981. This one is destined for Venezuela.*

ABOVE *Designed for customers with real STOL problems, the de Havilland Canada DHC-7 'Dash 7' is a rather specialized vehicle. In 1980, however, sales took off, largely because of the possibility of operating from taxiways or short runway-ends without disturbing the main traffic at major airports.*

This STOL category in the 12-to-50-seat size has become by far the most important of all civil turboprops. Despite numerous early projects the first to catch on to an important market was the de Havilland Canada DHC-6 Twin Otter. This Toronto-based company has made itself world-leader in STOL utility transports, of steadily increasing size and capability, with long wings liberally endowed with slats and flaps and large tails able to retain control at what in a stiff breeze looks like a mere walking pace.

The trim Twin Otter has two 580 horsepower Canadian Pratt & Whitney PT6 engines, a high wing, fixed landing gear and seats for up to 20 passengers. There are various versions, including those equipped with ski or float landing gear, and by late 1980 sales had reached 750. The next DHC was simply named the Dash-7, its designation naturally being DHC-7. Much larger, more complex and expensive than the Twin Otter, it is powered by four PT6 engines of a later model rated at 1,120 horsepower, geared down to drive large slow-turning propellers for minimum noise. With even more extreme high-lift slats, flaps and flight controls than the earlier DHC aircraft, the Dash-7 seats up to 54 passengers in a pressurized fuselage of nearly circular section, and is very much a specialized vehicle for operators afflicted with impossibly short airstrips, mountains, tropical temperatures, high-altitude airports or a combination of these problems. It also offers outstanding quietness, and therefore is ideal for services to short strips right in the heart of cities. The prototype flew on 27 March 1975, but for three years the sales could be counted on the fingers, almost, of one hand. Then, as is the way with airlines, operators suddenly fell over each other to buy Dash-7s in 1979 and sales zoomed close to 100 in months. Now the smaller DHC-8 is on the drawing board, seating 30 and powered by two of the brand-new Canadian Pratt & Whitney PT7 engines. High-winged, like previous DHC designs (but unlike most of its rivals), the Dash-8 was said in mid-1980 to have attracted 64 'letters of intent' from 19 customers, all of them commuter airlines or oil companies and, with the exception of Brymon in Britain and South Pacific Island Airways, all in North America.

Britain's contribution to this booming market began with the unpretentious Short Skyvan, first flown in 1963 and once described as 'a shed with wings'. Powered by two 715 horsepower Garrett TPE 331 engines, it has a high strut-braced wing, twin rectangular fins, fixed landing gear and an unpressurized fuselage with full-section rear loading for freight or small vehicles. The Skyliner is a more luxurious all-passenger version seating up to 22. From this the company developed the Model 330, an outstanding 30-seat commuter airliner with two 1,173 horsepower PT6 engines turning quiet five-blade propellers and with a comfortable big-windowed interior and retractable landing gear. First flight was on 22 August 1974 and by 1980 more than 30 were actually in use on four continents. Today the company has added an even better-looking 36-seater, the Short 360.

Another user of the ubiquitous PT6 engine, though at the 680 horsepower level, is Brazil's immensely successful EMB-110 Bandeirante, leading product of a diverse and prosperous company which calculates itself to be No 6 in the world in terms of aircraft produced per year. There are more than a dozen variations on the -110, but the basic passenger machine seats 15 (one model seats 21) and has even been exported to Britain and France. An American rival is the sleek but rather cramped Swearingen Metro, which has two 940 horsepower TPE 331 engines and seats up to 22 passengers in a shallow tube-like fuselage pressurized for high-altitude flight. Another machine in the same class is the Jetstream, originally produced by Handley Page in Britain and first flown in August 1967. After delivering 38 of these extremely popular 18-seaters the company collapsed. Following many vicissitudes the improved Jetstream 31 flew at the British Aerospace Scottish Division in March 1980 and is assured of very large orders not only from commuter and corporate (executive) customers but also from military services including the RAF and Royal Navy, both of which use earlier Jetstreams.

Czechoslovakia's Let L-410, powered by either two PT6 engines or the locally produced M601, is a 15- to 20-seater which, despite its rounded appearance, is unpressurized and basically simple. It beat a Russian machine for large orders by Aeroflot. An even simpler aircraft of superficially similar appearance, Spain's CASA 212 Aviocar has 776 horsepower TPE 331 engines and can seat 19 passengers or load vehicles or bulky freight through a rear ramp/door. Much smaller is Australia's GAF Nomad, with 400 horsepower Allison 250 engines, seating either 12 or, in a stretched model, 15. It is popular in third-world regions for transport, ambulance and various special duties, and can have twin-float landing gear.

West Germany's Dornier has made many STOL piston-engined utility transports and is now developing the advanced-technology twin-turboprop LTA in the 30-seat class, as well as the Model 228 to seat 15 (228-100) or 19 (228-200) on two TPE 331s, the prototype being due to fly in 1981. Gulfstream American, heir to Grumman's executive aircraft business (see next chapter) has taken the famed Gulfstream I and stretched it into the GAC-159C, still powered by two Darts, seating as many as 38 and with an optional cargo door. This attractive machine with old engines will be in head-on competition with a brand-new commuter airliner in the overcrowded 30-seat class to be developed jointly by Fairchild of the USA and Saab-Scania of Sweden. Its engines will be the new CT7-5 in the 1,700-horsepower class, and the first flight is due in 1982. In France, Aérospatiale is working on the AS 35, a high-winger which may have CT7 engines, and the company hopes to collaborate with Aeritalia (which has its own AIT 230 project). Now, such is the appeal of the propeller, Lockheed is even studying a turboprop version of the TriStar!

ABOVE *The Embraer EMB-110 Bandeirante is made in Brazil but sold all over the world, including Britain. Embraer is now building the larger (30-seat) EMB-120 Brasilia. Powered by Canadian Pratt & Whitney engines more than twice as powerful as the 680 hp units of the Bandeirante, the Brasilia is to fly in mid-1982.*

G-BBEE
N362GL

GENERAL AVIATION
Executive jets, air-taxis and helicopters

General aviation (GA) is the name for all civil aviation apart from the airlines. One of its largest sectors is agricultural flying, and another is private flying, but these are inappropriate to a book on airliners. It does, however, include a vast amount of transport aircraft of many kinds ranging right up to the mighty 747, operated by heads of state, all types of companies especially international industrial corporations, air-taxi and charter firms, and of course helicopters. Indeed, a few helicopters are operated by airlines, though even these are often wholly committed to special tasks such as the support of offshore oil platforms.

Private companies began to use aircraft long before this became commonplace. For example, immediately after World War 1 the sales manager of Boulton & Paul Ltd, of Norwich, used one of the company's own products, a P.6 two-seater, on business trips; and the 100th corporate aircraft in the United States was put into use as early as 1928. After World War 2 the number of executive aircraft in North America rose well into four figures. But not one was designed for the job. In the case of large corporations which needed to convey as many as a dozen employees over transcontinental distances the answer was either a secondhand airliner, a DC-3 or one of the speedy twin-engined Lockheeds, or a converted bomber.

During the 1950s such modern short-haul airliners as the Convair-Liner family (240, 340 and 440) and Viscount were sold not secondhand but new to major corporations. Limited sales were achieved by small twins such as Britain's Dragon Rapide biplane and Airspeed Consul and DH Dove monoplanes, and in France there were even small four-seat jets such as the Morane-Saulnier Paris. Then in 1955 Dassault, a French company specializing in jet fighters, announced something rather different – an executive jet. The Méditerranée was to be a swept-wing machine using the same wing as the Mystère fighter but with a pressurized fuselage seating up to eight, and with two Viper turbojets at

BELOW *Though by no means the most roomy executive aircraft, the Gates Learjet family has outsold all others, the total passing 1,000 in March 1980. The main reasons are the diversity of models available, their very high flight performance, and low price. Here a fan-engined Model 36 (in the foreground) is seen in company with the first British Model 25B.*

the tail. Like an executive version of the Martin B-57 (British Canberra) jet bomber, it was never built, but in concept it was right. Then the US Air Force issued a specification for a UCX (utility jet transport) and UTX (utility jet transport/trainer). The UCX led to the Lockheed JetStar, flown on 4 September 1957, and the UTX triggered off the North American (later Rockwell) Sabreliner, first flown on 16 September 1958.

Both found not only various military but also corporate markets. The JetStar remains the only purpose-designed four-engined business jet (bizjet), typically seating about ten in addition to two pilots and noteworthy for its big fuel pods on the wings. The Sabreliner has been developed in models with different wings, engines and fuselages, the original slim model with triangular windows having been succeeded by versions with almost square windows and more headroom.

In 1956 Grumman, a well-known builder of aircraft for the US Navy, boldly began the design of the G-159 Gulfstream, a business aircraft of extremely modern design powered by two of the trusty Rolls-Royce Dart turboprops. The rather slim fuselage provided comfortable accommodation for up to a dozen passengers, with wide horizontal-ellipse windows. From the first flight on 14 August 1958 this gamble proved a success and, though rather expensive, Grumman eventually sold 200. The last was delivered in 1969, but four years earlier Grumman had begun the design of a jet successor. This was named Gulfstream II, the original model becoming the G I. The first G II, which was the first off a busy production line, flew on 2 October 1966 and despite its high price found a ready market. Powered by two Rolls-Royce Spey turbofans, it has slightly more cabin space than the G I and much greater speed and range. From the start the G II has been the Rolls-Royce or Cadillac among bizjets, and it has found a market among governments and for corporate and crew-training use by airlines. The factory at Savannah, Georgia, became part of a new company, Gulfstream American, in 1978, and G II production ceased there in 1979 at No 256. But it was at once replaced by the refined G III, with additional fuel capacity and significant airframe changes including winglets, small extra upturned aerofoils at the wingtips which improve airflow and extend range. Some G IIIs are furnished for as many as 19 passengers, but ten is more common in VIP executive use with each interior styled individually.

Britain got into the act in 1962 when de Havilland (by then part of Hawker Siddeley) flew the Jet Dragon DH 125, which went into production as the Hawker Siddeley 125 and today is selling better than ever as the BAe (British Aerospace) 125. It scores by being a middle-of-the-road machine, offering more headroom and comfort than the fastest bizjets and able to use relatively short runways. The first batches were powered by the Rolls-Royce Viper turbojet, but in June 1976 the 125-700 took to the air with quieter and more fuel-efficient American TFE 731 turbofans which are also being retrofitted to a few earlier machines. By late 1977 sales were nudging 400 and today they have passed the 500 mark.

Best-selling of all the bizjets is the Learjet family. Bill Lear, one of the greatest names in general aviation, designed the original Learjet in Switzerland but it flew at Wichita in 1963 and was later taken over by Gates Rubber Company to become the Gates Learjet. Early versions were built like a Jaguar or Maserati, seating six passengers in great comfort provided they did not try to stand up. Flight performance was out of this world, and soon an eight-passenger model was added. Then the number of variants proliferated, some having quiet

BELOW *Leading producer of bizjets in Europe, Dassault Breguet of France offers three basic Falcon models, of which the biggest and the most expensive is the ultra-long-range Falcon 50. A trijet, with three Garrett TFE 731 turbofans, it has already notched up nearly 140 sales.*

fuel-efficient fan engines and the latest Series 54, 55 and 56 having 'stand-up' interiors for ten passengers and new wingletted wings with which they can zoom straight up to an unrivalled cruising level of 15,550 m (51,000 ft). Gates Learjet has been No 1 in bizjet deliveries for 13 consecutive years, the 1,000th being supplied in 1980.

The major producer outside the United States is Dassault-Breguet of France, whose original prototype, called the Mystère 20, flew on 4 May 1963. Thanks to favourable evaluation by Charles Lindbergh it was selected by Pan American, which wished to launch a Business Jets Division and, after various changes and being renamed the Fanjet Falcon, the French machine began to pour into the North American market, which is larger than all other countries combined. Powered by General Electric turbofans, the Falcon 20 has powered flight controls and typically seats ten passengers, but one enterprising operator, Federal Express, of Little Rock, Arkansas, uses no fewer than 33 in all-cargo configuration. Encourage by the success of the Falcon 20, Dassault launched a smaller aircraft, the 4/7-passenger Falcon 10, first flown on 1 December 1970, and then an advanced three-engined long-range aircraft with a supercritical wing, the Falcon 50, first flown with this wing on 6 May 1977 after testing with an older wing. By mid-1980 Dassault had sold 197 Falcon 10s, 468 Falcon 20s and even 128 of the rather specialized Falcon 50s.

In contrast, the nationalized rival, the great conglomerate called Aérospatiale, sold only 40 of its neat Corvettes, and did not launch a planned enlarged version with seating increased to 18 instead of 12. Neither did the West German company MBB have much success with the HFB 320 Hansa Jet, though this was a novel design with swept-forward wings. This shape, which demands very stiff wings in order to avoid aeroelastic (combined aerodynamic and structural) distortion, was adopted in order that the wing could be mounted in the mid position behind the rear cabin pressure bulkhead, thus leaving the interior completely unobstructed. Most other small jets have an annoying step in the cabin caused by the structural box of the wing centre section.

Another unsuccessful European small jet was the Piaggio-Douglas PD-808, a joint effort by Italy and the USA with British engines, notable for not having swept wings. However, another 'straight-wing' jet has sold well in many markets. Aero Commander was started in the immediate post war era by Ted Smith, a former designer of the Douglas A-20 (called Boston or Havoc), a famed World War 2 bomber. His Commander looked like a small version of the A-20 when it flew in 1949, and by the mid 1950s it was an established product with two 240 horsepower piston engines and seating up to seven. While various more powerful and faster models appeared a jet version was also designed, and this flew as the Aero Commander 1121 on 27 January 1963. It was an immediate success, with two General Electric CJ610 turbojets hung on the rear fuselage so that the long cabin could lie entirely ahead of the mid-mounted wing. However, in 1967 North American Aviation, which had acquired Aero Commander, was merged into Rockwell, and to avoid having two competing bizjets in one company the Sabreliner was kept and the Commander 1121 sold to Israel Aircraft Industries. IAI has since developed several new versions including the lengthened ten-seat Westwind 1123 and the Westwind 1124 with economical turbofan engines, and today the Lod (Lydda) based company has sold over 120 of this new family.

Until 1968 none of the giants of general aviation, Cessna, Beech and Piper, had touched the bizjet market. In that year the most prolific planemaker, Cessna – whose deliveries passed the 150,000 mark in 1978 – announced the Fanjet 500, soon renamed Citation I. First flown on 15 September 1969, this was the cheapest of all small twinjets, with a straight wing, two Pratt & Whitney of Canada JT15D turbofans on the rear fuselage, and eight seats including room for one or two pilots. In 1977, when 400 Citations had been sold (later examples having a greater span and more powerful engines and being styled Citation I), Cessna flew the first Citation II with a stretched fuselage seating up to 12, and extended range. Throughout the second half of the 1970s the company had been studying larger, longer-range versions far beyond the C II and, after detailed work on a straight-wing trijet with a T-tail, the totally new Citation III (C III) was launched in late 1976, the prototype flying on 30 May 1979. This still has only two engines but they are TFE 731s, much more powerful than the engines of earlier Citations, and the completely new 15-seat fuselage rides on a swept supercritical wing of advanced design. Range is similar to the G III and Falcon 50 at 5,550km (3,450miles).

ABOVE *Because it exists in many versions, the British Aerospace 125 is one of the world's bestselling executive jets, with 500 sales by late 1980. Production is now entirely of the fan-engined Series 700, which in addition to quieter and more fuel-efficient TFE 731 engines incorporates many airframe improvements. Sales of this new model already exceed 150.*

RIGHT *Cessna, the world's biggest lightplane producer, entered the executive jet market in 1968. After many changes the first Citation flew in 1969 and over 600 of the basic 5/6-seat model have been sold. In January 1977 the 7/11-seat Citation II made its first flight, and one of these is illustrated. The main distinguishing feature is that there are six windows on each side instead of four.*

BELOW *In May 1979 Cessna flew the completely redesigned and much more advanced Citation III, with almost twice the range of earlier models at 160 km/h (100 mph) higher speed. Deliveries begin in 1982.*

Another new bizjet in this class was designed by Bill Lear in the early 1970s as the LearStar 600, taken over by Canadair and much modified and finally put into production as the Challenger, the first example flying on 8 November 1978. From the start this has been a bold and ambitious project with two Avco Lycoming turbofans of much greater power than those fitted to any other bizjet except the Gulfstreams (they are the same as the engines of the BAe 146 airliner). They are hung on a fuselage of exceptional diameter, giving a cabin significantly larger than in other small twinjets, and the wing is one of the most advanced supercritical types to have flown. Most Challengers delivered from 1980 have about a dozen seats in all, but the even more powerful Challenger E (whose General Electric CF34 engines are a customer option on the original model) has a stretched fuselage and also greatly increased range (E stands for Extended-range). This bigger version flies in 1981.

So buoyant is the bizjet market that new manufacturers are still making their appearance. One of the most recent entrants is the Japanese giant Mitsubishi, whose MU-300 Diamond I first flew on 29 August 1978. Mitsubishi had previously sold well over 600 MU-2 twin-turboprops of strangely compact high-winged layout in various civil and military markets, but the Diamond is a straightforward machine following the almost universal formula for bizjets of a low wing and rear engines, the latter being JT15Ds from Montreal.

One of the smallest cabin aircraft ever built, and an interesting exercise in trying to tap a new market, the original Foxjet was planned in Minneapolis in 1977 as a true baby bizjet powered by Williams turbofans of a mere 360kg (800lb) thrust each, and with a pilot and three passengers shoehorned into the circular-section fuselage in tandem. This eventually appeared too cramped for real success and in 1979 more powerful Williams engines were chosen and the cabin was enlarged to 1.4m (55in) diameter overall; the Foxjet ST600-S/8 can now seat a pilot and five passengers. It looks fast even when standing still, and with aggressive marketing Foxjet expect to sell 44 a month.

Bill Lear, who died in 1978, left behind plans for a remarkable new business aircraft powered by a pusher propeller. Lear's long-range vision saw that corporations would eventually become more interested in jet noise and the price of fuel, and this inevitably led either to turbofans of very high bypass ratio or to the propeller. He chose the latter route, and in 1980 Learavia Corporation was well advanced with the Lear Fan 2100. Predictably its technology is almost out of this world, for the airframe has an almost perfectly streamlined fuselage, Y-shaped tail and an extremely advanced straight wing constructed in graphite/epoxy, boron fibre, glass fibre, Kevlar fibre and various resins. Nine seats, the front one or two of which are for pilots, fill a cabin of carefully optimized dimensions behind which are two Pratt & Whitney of Canada PT6B turboshaft engines. These are geared to a shaft extending to the extreme rear of the fuselage where, via a gearbox and clutches allowing either engine to continue to drive following failure of the other, they turn a slow-running reversible-pitch propeller. Most modern turboprops can drive reversible propellers, whose blades can be set to an angle at which they give not thrust but a

RIGHT *The only British lightplane of recent years to sustain production has been the successful family of Islander/Trislander light transports designed by Britten-Norman in the Isle of Wight. Following a 1978 bankruptcy, despite a soaring order book, the company became Swiss-owned as Pilatus Britten-Norman. The photograph shows Islanders and (nearest the camera) Trislanders in final assembly at Gosselies, Belgium. This was terminated, and assembly now takes place in the Isle of Wight, Romania and the Philippines. Over 1,000 of these versatile aircraft have been delivered.*

INSET *A stretched Islander with a third engine on the fin, the Trislander seats 18 instead of ten, including the pilot.*

G-BEKY

ABOVE *Large numbers of speedy twin-turboprops are produced by Swearingen Aviation of San Antonio, Texas, a subsidiary of Fairchild. This is a Merlin IIIB, an 8/11-seater, one of which was delivered in 1979 non-stop from New Jersey, USA to Cologne, Germany.*

braking force, corresponding to the action of a jet engine with the reverser in operation. The Lear Fan is expected to cruise at 640 km/h (400 mph) and give unrivalled all-round performance and economy.

Another unconventional aircraft in this class is the Gulfstream American Hustler 500. The prototype, designated Hustler 400, flew on 11 January 1978 and introduced another novel way of reconciling the conflicting demands of high performance, comfort, long range and economy. The only normal feature is the pressurized fuselage with two crew in a forward cockpit and up to seven passengers in the main cabin. The Hustler 400 is powered by a PT6A turboprop in the nose and has a T-tail, the tailplane being on top of the fin. But the production model, the Hustler 500, also has a JT15D turbofan in the rear fuselage, fed through a flush NASA-type vee-shaped inlet immediately ahead of the fin. The nose engine has been changed to a Garrett TPE 331 because this can have a single exhaust pipe underneath, which minimizes ingestion of hot gas by the rear engine. The wing is of the supercritical type and has high-lift flaps for short field performance. As originally conceived the Hustler was to be a turboprop with a small auxiliary rear jet for take-off and emergencies, but now it is seen as a twin-engined aircraft which will cruise on both engines at 741 km/h (461 mph).

In a book about airliners not too much need be said about ordinary single-engined general aviation aircraft, though these make up by far the greatest category in the total GA market of some 16,000 aircraft a year in Western countries. But the group generally known as 'light twins' has grown tremendously. Many of these machines have power, seating and range similar to those of typical airliners of the 1930s, and about twice the speed. Typical examples of these six/11-seaters are the Cessna Titan, Piper Navajo and Beech Baron, while more powerful turboprop models are the Cessna Conquest, Piper Cheyenne and the big Beech Super King Air which can accommodate 13 (in 1981 Beech expected to deliver the first 19-seat Commuter). Most belong to businesses, taxi and charter firms which ply for hire at virtually all Western airports and will fly anywhere safe and legal.

Special mention should be made of Britain's Pilatus Britten-Norman Islander and Trislander. First flown on 13 June 1965, the Islander was planned as the simplest possible transport seating more than six passengers or well over a tonne of cargo. It came out as a ten-seater with so many convenient doors that there is no need for an aisle, which means the fuselage can be slimmer. With twin main wheels for soft soil or sand, Islanders go

anywhere an aeroplane can fly, and by 1980 more than 1,000 had been delivered for a vast range of duties. The Trislander is a neat enlarged version with a fuselage stretched to seat 17 passengers and with a third engine uniquely added at the junction of the fin and the high-set tailplane.

Such machines can use airports almost as small as a major heliport. In the early 1960s there was much enthusiastic discussion of V/STOL (vertical or short take-off and landing) airliners, which were thought to be the solution to airport overcrowding, air-traffic control delays and the inconvenience of the often long journey from the city to the airport. However they failed to catch on and the only city-centre machines continue to be helicopters, and even here the giant airliner helicopters have been a very long time coming.

Scheduled helicopter services with mail were flown by British European Airways in 1950, and Sabena (Belgian airlines) sustained passenger services between several north-west European cities for much of the 1950s, but it was not until the turboshaft engine gave the helicopter extra power and reliability in the late 1950s that airline services slowly began to 'take off'. Los Angeles Airways and New York Airways used twin-turbine helicopters on busy passenger and mail networks throughout both these extremely large and prosperous cities, and in the 1960s business helicopters such as the Bell JetRanger began to swarm all over the world. A large segment of the market was accounted for by the world oil industry which finds the helicopter useful because of its ability to inspect pipelines and service oil rigs. One oil support company has a fleet of 135 JetRangers, and the North Sea platforms are serviced by hundreds of helicopters of many kinds.

In 1981 British Airways Helicopters was to begin operations with the biggest and most capable airline helicopter yet, the Boeing Vertol 234. Derived from a mass-produced military helicopter of the tandem-rotor type, with front and rear rotors geared together and driven by two 4,353 horsepower engines at the tail, the Model 234 has a main cabin not much smaller than that of typical short-haul fixed-wing airliners, seating 44 in double seats each side of a central aisle. In what is called a typical 'combi' configuration (a word used for all airliners equipped for the mixed passenger/freight role) it can carry 18 passengers and 7,250 kg (16,000 lb) of cargo. Though the four landing gears do not retract it has a useful turn of speed, cruising at 263 km/h (163 mph) and like all airline helicopters has full all-weather equipment.

In fact there are even larger and more powerful helicopters, but these are of the heavy lift or crane type used for lifting and positioning large loads in isolated or difficult places. Typical tasks include placing tops on TV masts, air-conditioning systems on the roofs of tall buildings, steeples on churches and electricity pylons across mountains.

ABOVE *Chief manufacturer of large airline helicopters in the West, Sikorsky Aircraft was also the original pioneer of the production helicopter. This S-61N of KLM Royal Dutch Airlines is one of 136 of this family of 30-seaters built in 1961-80.*

OPPOSITE *By far the best-selling among light turbine helicopters, the Bell JetRanger is sold in various models typically seating 5/7 passengers or two stretcher patients and two attendants. Over 7,000 of all models have been produced.*

SUPERSONIC TRAVEL AND THE FUTURE

BELOW *A fantastic technical success, but hardly a commercial one, Concorde (shown here in Air France livery) has suffered from not being American. This has restricted its market, and operations have been further hit by escalating fuel prices and political interference which in November 1980 forced British Airways to close services to the East.*

The first human to fly faster than sound was Charles 'Chuck' Yeager, then a captain in the US Army Air Force, on 14 October 1947. He was in the needle nose of a special rocket-propelled research aeroplane, the Bell XS-1, and he had to be strapped tightly because at around the speed of sound the XS-1 behaved like a bucking bronco, trying to break its wings and smash its pilot's head against the canopy. Nobody knew much about airflow at such speeds, nor about how to make supersonic aircraft, except that they needed the maximum amount of power and the minimum of frontal area.

Gradually knowledge was gained, at considerable cost in money and, sometimes, in lives. By the mid-1950s fighter aircraft were in regular use that ordinary squadron pilots could fly faster than Mach 1 (Mach 1 is the local speed of sound, which varies according to air temperature so that it is greater on a hot day, and much greater at sea level than at high jet altitudes). Test pilots at the same time were flying fighters able to reach Mach 1.5 or even 2, and in 1956 testing began of a bomber that could hold Mach 2 for about an hour. This impressive aircraft, the B-58, was naturally studied as a

possible basis for the world's first SST (supersonic transport), but though General Dynamics, the builder, produced some rather attractive brochures the airlines of the world were far from ready for a supersonic airliner.

Another event of 1956, little known even among aviation people, was the first meeting of the British STAC (Supersonic Transport Aircraft Committee). This included representatives from manufacturers, engine companies, airlines and government organizations, and spent years carefully looking at the possibilities. There appeared to be three choices of cruising speed. A light-alloy aircraft with a curious kinked M-shaped wing could cruise at about Mach 1.2 (20 per cent faster than sound, about 1,275 km/h [792 mph] at high altitude) and, though like all supersonic bodies it would have to generate shockwaves around it, these would never reach the ground because of the increasing air temperature at lower levels and the consequent curvature of the shock fronts. This was expected to be a social bonus, but the aircraft was otherwise unattractive. An SST cruising at about Mach 2.2 (2,337 km/h or 1,452 mph at high altitude) could

still be built in light alloy, though the metal would have to be specially heat resistant. The limit of technology was a Mach 3 (3,220 km/h or 2,000 mph) airliner, which would have to be made of steel and titanium to withstand the kinetic heating of sustained flight at such speed.

It so happened that the best engine efficiency would be gained at Mach 3, because of the tremendous compression of the airflow entering the engine inlets. But extreme difficulty and resulting delay, and soaring costs, with a stainless-steel research aircraft, the Bristol 188, indicated that a go-ahead on so challenging a proposal would be bitterly regretted. Therefore the decision was cast in favour of Mach 2.2. Bristol Aircraft did intensive research and came up with a tailless delta shape which did not have the triangular wing of the true delta but a very long and gracefully curving form called an ogival or 'gothic delta' outline. Work went ahead on the big Bristol 223 for carrying 110 passengers between London and New York at Mach 2.2.

By the 1961 Paris air show the 223 design had reached an advanced stage, when the British designers were intrigued to see something that looked extremely like it in the form of a model on the French Sud-Aviation (later renamed Aérospatiale) stand. Called Super Caravelle, the model did not represent an actual design but merely what the French company thought would be a logical successor to its Caravelle. Like that famous jetliner it was a short-haul transport, with nothing like enough range to cross the Atlantic. But it seemed a good idea to the British government to get the two design teams together and at ministerial level hammer out a joint collaborative programme. The two companies began discussions as early as 8 June 1961, during the Paris air show, but an intergovernment agreement was not signed until 29 November 1962. By 1963 the aircraft was first drawn on paper and named Concorde.

One helpful fact was that the basis for the engine was already well proven: Bristol, later to be Bristol Siddeley and from 1966 Rolls-Royce, had fully developed a large and efficient turbojet, the Olympus, for the Vulcan subsonic bomber and was already well advanced with a much more powerful supersonic version with an afterburner (a special jetpipe in which additional fuel is burned to give extra thrust) for the TSR.2 supersonic bomber. To assist the desired 50/50 share of the work the afterburner and nozzle of the engine were handed to the French engine company SNECMA. In order to give the pilots adequate view for take-off and landing the entire nose was arranged to hinge down, in what the media called a droop-snoot. Lift at low speeds was greatly enhanced by the design of the sharp-edged wing which at large angles, as at take-off, was arranged to form a powerful writhing vortex of air which curves back across the wing and makes the latter appear to the airflow much thicker and with a generous rounded leading edge, like a high-lift wing on subsonic aircraft. On the other hand a tailless delta, like Concorde, has to take-off and land faster because, instead of having high-lift flaps on the trailing edge, it has large 'elevon' (combined elevator and aileron) control surfaces which, to raise the nose, press downwards on the trailing edge and have the effect of adding weight.

To preserve the image of a 50/50 programme a separate production line was set up in both countries, though there was no actual duplication of manufacture, for each component was made in just one location. The French assembly line was at Toulouse, and here Concorde prototype 001 flew on 2 March 1969. The British assembly line was at Filton, north of Bristol, and here 002 followed on 9 April 1969. These machines had many interim features and a gross weight of 147,870 kg (326,000 lb). Then followed pre-production aircraft 01 (Filton) and 02 (Toulouse), which had longer fuselages and a pressurized interior lengthened by no less than 5.9 m (19 ft 3½ in). Other changes concerned the engines, wings and vizor, the latter giving a dramatically different nose appearance. Finally the production machines numbered 201 to 216 appeared between 1973 and 1979, with a much longer rear fuselage projecting far behind the trailing edge of the rudder, greater gross weight (up from 174,635 to 185,065 kg/385,000 to 407,994 lb) and engines with improved variable nozzles.

Curiously, this civil transport vehicle became the target of protests of intense violence from people who claimed it would harm the environment. These protests and prolonged legal proceed-

BELOW *No airliner in history has suffered more prolonged development nor passed through greater changes than the Soviet Union's Tu-144. Rather larger and more powerful than Concorde, it is pictured at Paris in 1975 in its original production form with two separate engine boxes into which the main gears retract (in the original aircraft the gears had three pairs of double wheels and retracted into the wing outboard of the single box for the four engines). Another added feature, dating from 1972, is the canard foreplane, seen here extended for take-off/landing along with the drooped nose. Sporadic services with Aeroflot were abandoned by June 1978, but a year later the much revised Tu-144D began further development with new engines.*

ings for years held up Concorde's admission to the United States, and particularly to New York's Kennedy Airport. Scheduled operations by Air France between Paris and Rio and by British Airways between London and Bahrein (as the first sector in a route to Australia and other Far East cities) began on 24 May 1976, and expanded to take in such cities as Washington, Mexico City and Caracas. Later Braniff of the USA flew Washington/Dallas and Singapore Airlines flew the Far East route in partnership with British Airways. Passenger load factor, the percentage of available seats actually filled, has been consistently high despite the imposition of a fares tariff higher even than normal First Class. Additional developments have cut fuel consumption by some 2,100 kg (4,630 lb) on long sectors, making this weight available for extra payload.

In contrast, the other SSTs have failed to show any return for their considerable costs. The first SST in the world to fly was in fact not Concorde but the Tu-144, created to an Aeroflot specification by the Soviet Union's chief design bureau for large aircraft headed by Dr Alexei N. Tupolev. The first prototype flew on the last day of 1968 and showed an almost complete similarity to the design of Concorde apart from having all four engines located in a single giant box under the wing. Fractionally larger than the Anglo-French SST, the Tu-144 was more powerful and was intended to carry rather more passengers (140) over greater distances. However, by 1973 the Tu-144 had been redesigned, with separate twin-engine nacelles and with retractable canard surfaces on the forward fuselage to lift the nose at low speeds and allow the trailing-edge elevons to deflect down (for lift) instead of up at take-off and landing.

The first of the redesigned aircraft crashed at Paris in June 1973, but after prolonged further testing one Tu-144 went into regular cargo and mail service with Aeroflot on 26 December 1975. Passenger service at last began between Moscow

LEFT *One of a profusion of published studies for future SSTs in the United States, this Lockheed LH_2-fuelled proposal is based on shapes refined in aerodynamic testing by NASA, the US national aerospace research agency.*

BELOW *Proposed by McDonnell Douglas, this HST (hypersonic transport) is LH_2-fuelled and intended to cruise at Mach 6, or 6,370 km/h (3,960 mph).*

and Alma-Ata, Kazakhstan, on 1 November 1977, but it was erratic and, following an accident to a Tu-144 not carrying passengers, was suspended on 1 June 1978. By this time the improved Tu-144D was well advanced in development, with quieter and more fuel-efficient engines, and this may eventually go into service.

In the United States the bold decision was taken to go for a Mach-3 SST, and following an industry competition in 1963-66 Boeing was selected to build an aircraft much larger, as well as faster, than the European rival. General Electric produced the GE4 turbojet, with afterburning thrust of 30,390 kg (67,000 lb), greater than for any previous aircraft engine. At first the Boeing 2707 aircraft had swinging (variable-sweep, pivoted) wings, and the engines were mounted under the large tailplane. In 1968 this was abandoned and the 2707-300 was decided upon, with fixed gull wings (gull wings curve up from the roots and then slope down to the tips) and the engines mounted under the wings in four separate nacelles. Typical seating was for 234 passengers, with structure almost entirely of titanium and steel, but after various environmentalist and political pressures the programme was abandoned in 1971.

After this debacle the American industry said little about SSTs for several years, but in the late 1970s once more tentatively began 'testing the market' with a succession of proposals. Throughout, the great NASA (National Aeronautics and Space Administration) has been quietly testing wind-tunnel models and advanced propulsion systems, and today the significant progress that has been made is causing a deep surge of opinion within the manufacturing industry that a new SST programme should be launched. Most of the studies, particularly those by Lockheed and McDonnell Douglas, feature a wing rather different from that of Concorde and essentially resembling a normal subsonic wing that has been greatly extended in chord (width from leading to trailing edge) so that its area is multiplied by about three. In all cases a horizontal tail is added, so that at low speeds (ie, take-off and landing) the wing can have high-lift leading-edge slats or droops and trailing-edge flaps (which are denied to Concorde).

Post-Concorde SST studies have also been made by the Anglo-French partners, now known as British Aerospace and Aérospatiale. These are known as AST (Advanced Supersonic Transport) studies, and could lead to an aircraft carrying 230 passengers over ranges of 8,000 km (5,000 miles) with efficiency higher than Concorde and with turbofan derivatives of the Concorde's Olympus engine giving greater power for much reduced noise. It is unlikely the work will progress beyond 'paper' studies because of lack of funding, so future SSTs are almost certain to be American or Russian.

An even more important area of research concerns high-subsonic aircraft, airliners designed to cruise at speeds a little below that of sound. Already the latest products from Boeing, Lockheed, McDonnell Douglas and Airbus Industrie are dramatically superior in many ways to those at the start of the wide-body era around 1970. Structures are incorporating new materials, such as advanced composites reinforced by graphite or carbon fibres, Kevlar or boron filaments. Active controls, in

BELOW *Since the mid-1970s many manufacturers, notably the US 'Big Three' of Boeing, Lockheed and McDonnell Douglas, have been studying large freighters tailored to the standard 8 by 8 by 40 (foot) container. This is one of many suggestions in which the containers fit in an almost all-wing aircraft, in this case with engines augmenting lift by blowing across the top of the wing.*

which the control surfaces are deflected automatically to reduce stresses in the structure, are combining with the basic idea of the flattish-topped, bulged-underside supercritical wing to give wings that are much more efficient than those of early wide-bodies.

Avenues towards future development are broad and diverse. Perhaps the most important single pathway is towards fuels that are not derived from petroleum, which is soon to become scarce and will continue to escalate in price. The long-term answer will almost certainly prove to be a gradual switch to LH_2 (liquid hydrogen), but this is fraught with difficulty and will cost many billions to implement. There is abundant hydrogen, for example in the water of the ocean; but it is so much less dense than today's hydrocarbon fuels that airliners will become bloated by swollen fuel tanks, either at front and rear of the passenger cabin, or below the passenger floor, or in giant pods on the outer wings. Lockheed has led in published LH_2 studies, and, on balance, favours putting the tanks at front and rear in a monster fuselage, despite the fact that the intensely cold liquid would be extremely dangerous to humans in a forced landing that ruptured the tanks.

Much interesting study has been done to explore new shapes for airliners, some of them stemming from research conducted in military programmes. For example a search for a new STOL freighter for the US Air Force led to upper-surface blowing (USB) which is flying in the Soviet An-72 civil prototypes and could even be applied to large cargo or passenger aircraft of nearly all-wing design with the load carried inside a wing carrying the engines close above the leading edge. One of the more conventional research aircraft in this class is to fly in Japan in 1982; this QSTOL (Quiet STOL) aircraft does have a fuselage, whereas Douglas and Boeing studies usually slim down the fuselage into a thin boom just to carry the tail.

For cruising at about Mach 1.2, leaving no rumbling sonic boom on the ground, tomorrow's designers have the slew-wing, which is much better than the M-wing of 1958. A relatively traditional wing is pivoted at its mid-point as one unit to the fuselage of the airliner, which takes off with the wing set at 90°, running directly from side to side. As it climbs and accelerates the aircraft is gradually changed in shape by pivoting the wing round until it is slewed at a sharp angle of about 60°. In this configuration it would be possible to cruise about 50 per cent faster than today's jetliners but with no perceptible sonic boom on the ground and various other advantages. NASA is testing the idea with a small twin-jet research aircraft, the AD-1, first flown in 1979.

In conclusion, today's engineers are finding that the airlines are so bothered by soaring fuel bills that they are prepared to listen to any truthful proposal that will significantly reduce fuel costs, even if it involves propellers. But there is one neat propulsion system that bridges the gap between the propeller and the jet. Dowty Rotol in England has since 1977 been testing ducted propulsors which are like small-diameter multi-blade propellers running inside a profiled duct. Visually they are reminiscent of modern high bypass ratio turbofans, but of course they can be driven by turbines or piston engines. Their most immediately evident attribute is amazing quietness with a low hum in place of the snarl, rasp or threshing vibration of ordinary propellers, but they also significantly increase flight performance. Unlike ordinary turbofan engines their blades can be reversed in pitch to slow the aircraft after landing. For speeds up to 800 km/h (500 mph) we shall probably see many ducted propulsors.

Indeed, the airship – for 30 years virtually ignored – came back into prominence in about 1970. Many enthusiasts have sought to show that quiet load carriers, cruising at about 160 km/h (100 mph) with cargoes of freight containers, could be helpful in air transport. Some believe they could hover over a factory, pick up a load and deliver it to a distant city centre or mountain top. One day this dream could be fact.

FOOT OF PAGE *Though old in basic concept the pivoted 'slew wing' has only recently attracted much attention, and in December 1979 a NASA-sponsored research aircraft, the AD-1, began exploring the principle. This illustration shows how a future slew-wing SST might look.*

BELOW *At first sight looking like an Islander that is trying to become a 747, this small testbed aircraft is actually one of great significance. Retaining its original piston engines, it has Dowty-Rotol ducted propulsors giving higher flight performance than traditional propellers, with much less noise. Gradually the ducted propulsor may become more common than the propeller.*

INDEX

References to illustrations are in italics.

Acknowledgements

The publishers wish to thank the following organizations and individuals for their kind permission to reproduce the photographs in this book:
Air France 74-75; Air Portraits Colour Library 32-33, 64-65, 76-77; Alitalia 38-39; Dr Alan Beaumont 9; Peter J. Bish 32 above, 33 right, 40-41, 48-49, 50-51 above, 54-55, 56, 66, 70-71; Boeing Aircraft Corporation 18-19, 50; British Aerospace, Hatfield 11 above; British Aerospace, London (MARS) 44-45; British Airways 20, 27 above; British Airways (MARS) 42-43; British Caledonian 36-37; Cessna Aircraft Corporation 68-69; CFM International 14; Richard Cooke 4-5, 28-29; De Havilland, Canada 62 below; Fairchild 72 above; Flack/API (Halcyon) 50-51 below; Fokker-VFW, Amsterdam (MARS) 16; Jan Freis/Image Bank 1; Mike Hooks 59 inset, 60-61, 71 inset, 72 below, 73, 78-79; K.P. Lawrence 10, 55 below, 58-59; Lockheed Corporation 8, 25, 77 above inset; Lufthansa Bildarchiv 24-25, 27 below; McDonnell Douglas Corporation 2-3, 46-47, 77 below inset; McDonnell Douglas Corporation/John Stroud 6-7; R.A. Nicholls/Halcyon 15, 17, 45, 57, 62 above, 63, 67, 79; Novosti Press Agency 52-53; Gabe Palmer/Image Bank 30; Pan American 35 inset; Saudi-Arabian Airlines/Interlink Advertising Ltd 28; Skinkis/API (Halcyon) 55 right; Michael Skott/Image Bank 12-13; G.M. Smith/Image Bank 22-23; Alan Timbrell, BAA Gatwick 21; United Airlines 42 inset; C.R. Ware/Halcyon 34-35; S. Wolf/Halcyon 10-11.
Endpaper artwork by Frank Kennard.

PDO 82-1033

Gh
Thief

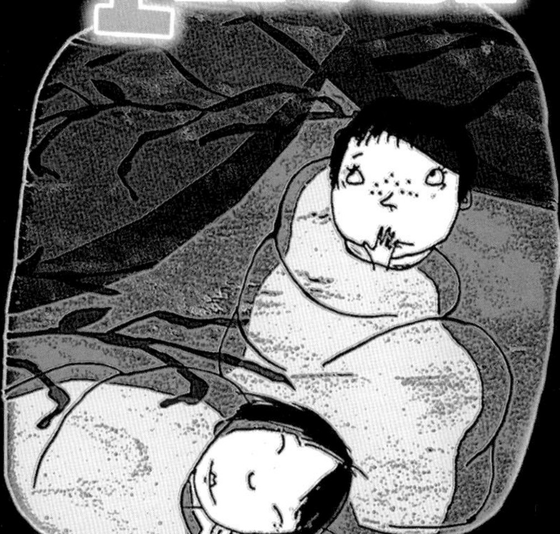

Kathryn White

Illustrated by Ella Okstad

Ginn

Chapter 1
A Ghostly Visit

Amy was in a bad mood. She hated being stuck in a tent with Ling. Amy glared at Ling, who was putting a pink, lace hair band in her black hair. "Yuck," thought Amy. "Ling is so girlie!" Amy liked fishing for newts, riding her BMX and *not* wearing silly hair bands.

Amy's mum popped her head through the tent door and gave the girls some leaflets. "Cheer up, you're on holiday now. Look – these tell you about the holiday camp," she said.

Amy took one look and sighed. "Boring!"

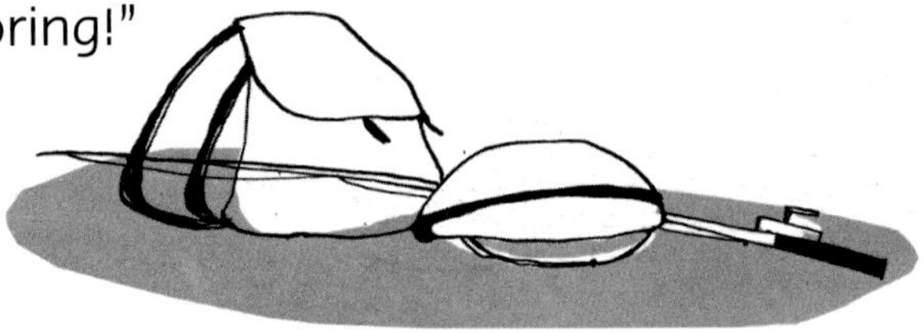

Ling was feeling just the same way because she wanted to be back home with her best friend, Lisa – not sharing a tent with moody Amy. Amy and Ling's mums were good friends and had decided to come on holiday together. But that didn't mean Amy and Ling were going to get on.

Ling opened a leaflet in a huff. "It says there are owls in the woods," she said, without much interest.

"Huh," said Amy. "Mine says there are squirrels."

Then Amy's eyes lit up. "A ghost!" she yelped. "There's a ghost here!"

Ling wasn't sure if she liked the sound of that or not.

Amy frowned and decided she wasn't keen on ghosts either. But she wasn't going to let Ling know this. "I'm not scared of ghosts," said Amy.

"Well I don't believe in them," said Ling, trying to sound cool.

"It says this ghost haunts the lake," said Amy.

"What kind of ghost is it?" asked Ling, quietly.

"A ghost boy. He went fishing on the lake and never came back." Amy held up a leaflet, which showed a picture of a boy with dark, curly hair. "He died one hundred years ago," said Amy, reading more from the leaflet.

"You're stupid if you believe all that!" said Ling, but as she looked over at the tent where their mums would be sleeping, she didn't feel so sure of her words.

That night, Ling and Amy lay sulking in their sleeping bags before falling asleep.

In the middle of the night, Amy was suddenly woken by an owl's hooting. She felt odd waking up in a strange place, and the ground felt hard under her. She could hear frogs croaking and lots of odd noises coming from outside the tent. Amy thought it was weird that they were lying under the moon and stars. There was no roof, no walls keeping out the night – just thin tent material.

The owl went silent. Amy turned over to go back to sleep but she heard something else. It sounded much closer to the tent.

Amy became alert. She lay still and listened as trees creaked and soft breezes made the leaves rustle. But this new noise was scary.

Something was tugging at the tent.

Amy froze. She held her breath.

Whatever was making the noise was only just on the other side of the canvas. If Amy reached out, she might even have touched it. The 'thing' was trying to get in. It moved around the outside of the tent and began to pull at the flaps. First it tugged on the door flap, then in a second it was pulling on the roof.

A horrible thought crept into Amy's mind. Whatever it was, it moved so fast that she couldn't even see a shadow.

There were only two things Amy knew that didn't have shadows – vampires and ghosts. What if it were the ghost boy?

The 'thing' shook one of the tent ropes. The top of the tent began to rock backwards and forwards, like a ship on the waves.

Amy sat up in her sleeping bag. "Ling? Are you awake?" said Amy, scared.

Ling didn't answer.

Then there was a terrible noise, like a long claw ripping down the side of the tent.

"Ling! Did you hear that?" cried Amy.

Just as Amy reached over to shake Ling awake, something fell down on the girls, covering them with a black cloak. They were trapped in the darkness.

Amy and Ling both screamed. They fought their way out of their sleeping bags, rolling into each other as they struggled to get free.

Just then, something grabbed Amy's foot. "Help! The ghost's got me!" she screamed.

"Argghh! I can't see!" said Ling, letting go of Amy's foot.

Finally the two girls crawled out from the collapsed tent. Their mums had heard the screams and were running over from their own tent.

"What happened?" said Amy's mum.

"It was a ghost!" yelled Amy.

Ling ran over to her own mum. "Amy's seen a ghost," she said, frightened.

"Calm down," said Ling's mum gently. "It sounds like you've both scared each other with silly ghost stories."

Amy's mum shone the torch around the girls' tent. It had caved in and the canvas was flapping in the wind. "Look, there's nothing here," she said. "One of the ropes has snapped, that's all."

Amy shivered beside her mum. "No, it hasn't snapped. I heard someone. They were trying to undo the rope," she said.

"But look, Amy. The rope was weak. It's ragged," said her mum. "You can both sleep in our tent tonight and we'll sort this out."

Ling and Amy ran into the other tent, and huddled in the sleeping bags. But they couldn't settle or even think of sleep.

Ling's face was white. "What do you think it was?" she said quietly.

"It was horrible. Something was outside. First it was in one place, then in another. It had no shadow, so I'm sure it was a ghost," said Amy.

"You're just trying to frighten me, aren't you?" said Ling sharply.

"No I'm not!" said Amy. "Anyway, I thought you said you didn't believe in ghosts."

Ling curled up into a tight ball and shivered. "I don't," she said quickly. "You're being stupid. It would have just been the trees rubbing on the rope. I never heard anything until you screamed."

But both Ling and Amy lay awake for a long time, just in case.

Chapter 2
A Mystery Begins

The next morning, Ling's mum put the tent back up. Then she called out, "Come on, sleepy heads. It's time to go and have a proper look around the holiday camp."

Ling was the first to crawl out into the sunshine, and headed to the other tent to find her clothes. Ling found her T-shirt and jeans, but couldn't find her pink hair band anywhere. It had gone. Ling hunted high and low for it, but it wasn't in the tent, and their mums hadn't seen it either.

With a face like thunder, Ling marched over to Amy, who was eating breakfast.

"Give me back my hair band," said Ling.

"What hair band?" said Amy, shocked.

“You know what hair band. I saw you making stupid faces at me yesterday when I was wearing it,” yelled Ling.

"Hey, what's the fuss about, you two?" said Amy's mum.

"She's taken my hair band," said Ling angrily.

"I did not. Why would I want a stupid thing like that?" Amy shouted back.

"Ling, you have plenty of other hair bands in your bag. Use another one for now and I'm sure it'll turn up later," said Ling's mum.

Ling scowled. "We won't find it because she's hidden it."

Amy stuck out her tongue. "It wasn't me," she called, as Ling stormed off. Then a worrying thought came into Amy's mind. If she hadn't taken the hair band then who had?

After breakfast, everyone went on a tour of the holiday camp. Amy smiled with delight when she found a big adventure play area. There were tube slides much bigger than any she had been on before.

After the play area, the girls were taken to the start of the nature trail. Ling wasn't happy about going. She had seen on a map that the nature trail went around the lake.

"I want to go to the caves instead," moaned Ling.

"The lake is on the way to the caves," said her mum.

"But I don't want to go to the lake," said Ling.

"What's the matter? Scared?" said Amy, with a grin.

"Never!" hissed Ling.

So, they set off on the nature trail and soon reached the lake. It was full of wildlife. Ducks scuttled over, eager for bread. A coot shot out from behind some reeds, followed by three tiny chicks. Two adult swans sailed by.

Much to Ling's surprise, the lake was full of wonderful things. Her mum gave her a bag of bread and Ling tossed handfuls to the birds. She felt much happier.

But Amy wasn't so happy. She was still worried about who had stolen Ling's hair band. She had a feeling that something strange was going on. Amy moodily skimmed a stone across the water. It bounced five times before sinking with a plop.

Ling smiled smugly. "You and your stupid ghost story," she said to Amy.

Amy prickled. "Watch out, Ling. It wasn't my hair band that the ghost took. It was yours!" she replied.

"Just give it back," snapped Ling, as she went back to feeding the birds.

That night, Ling switched off her torch and snuggled down to sleep. Amy was already in her sleeping bag pretending to be asleep. Neither girl had spoken much to each other all day.

Amy finally drifted off to sleep, but woke in the night again. She felt a cold hand touch her cheek. Something was close to her face. Sleepily, Amy tried to flick it away. Then she opened her eyes wide and looked at Ling. Ling was snoring loudly. But something *was* in the tent with them.

Amy watched in terror as Ling's shiny watch moved along the ground and vanished under the side of the tent.

Amy quickly leaned over and shook Ling awake.

"What's going on?" Ling asked sleepily.

"The g ... g ... ghost. It stole your watch," stammered Amy.

Ling could see how scared Amy was. She flashed her torch around the tent.

No one was there. "You're lying again," said Ling angrily.

"I'm not! I saw your watch just vanish!" cried Amy.

Ling hunted around for her watch. She couldn't find it. "You've hidden it," Ling snapped, rubbing her eyes. "You're not going to scare me!"

"I did not hide it. Honest," said Amy.

But Ling didn't want to listen. She turned off her torch and moved further away from Amy. "I'll find my things in the morning," hissed Ling, "and when I do ... you'll be sorry."

Amy was now scared and angry. She knew the truth. She knew that there was a ghost in the holiday camp.

The next day, Amy and Ling were allowed to go off to the adventure play area on their own.

"You know you must stick together," said Ling's mum.

"And be back by lunchtime," added Amy's mum.

The girls walked down the path without speaking to each other. When they reached the play area, Amy ran for the tube slides and Ling played on the tyre swings. They ignored each other all morning.

Just as Ling was about to go on the climbing frame, Amy raced up to her. "It's one o'clock. We've got to go back now," she said, looking at her watch.

"When we get back to the tents, you'd better show me where you hid *my* watch," snapped Ling.

"I've already told you. The ghost boy took it!" said Amy.

"I don't believe you. You won't scare me, Amy, and to prove it, I'm going back to our campsite on the path by the lake," said Ling. She immediately wished she hadn't said it.

Amy went white. "You can't do that! What if the ghost boy comes?" she cried.

"I'm not stupid," said Ling, already marching off down the path towards the lake. "You're just a liar."

Amy watched Ling disappear between the trees. She wanted to run back to their campsite, but their mums would be furious. They had been told to stick together. Amy had no choice. She followed Ling down to the haunted lake.

Chapter 3
The Haunted Lake

Ling was halfway round the lake when Amy saw her on the path. Amy's knees went weak and she gasped. Across from Ling was the ghost boy! He was standing out in the lake. The ghost boy was floating on top of the water! Amy knew she was seeing the ghost boy from the leaflet. She screamed in horror.

Ling heard Amy's scream from the other side of the lake. She turned to see Amy waving to her.

"Run, Ling, run," shouted Amy.

"Get lost," cried Ling. "You can't scare me, Amy."

"But there's a ghost on the lake!" shouted Amy.

Ling now looked out onto the water and her face turned to stone. An icy chill ran down her back. Amy was right. The ghost boy was there, floating on the water.

Ling yelled in terror and ran as fast as she could, back along the path towards Amy. As Ling reached Amy, they both looked back and saw the ghost boy dive into the water and vanish.

"Run faster," called Amy.

They raced back to their campsite so fast that when they stopped they could hardly speak.

Amy bent down, gasping for air.

"I'm really sorry I didn't believe you," gasped Ling.

Amy suddenly felt happy. At last she wasn't the only one who knew the truth about the ghost boy. "It's okay," said Amy.

Ling shuddered. "Why has the ghost boy taken my things?" she asked.

"I think I know," said Amy, wondering if anything else had been stolen. "We'd better check to see what's missing."

Before lunch, the girls hunted through their tent.

Amy yelped in horror.

"What is it?" asked Ling.

"My beads have gone," Amy said, "and one of my stripy socks."

Ling looked puzzled. "Why would a ghost want a sock?" she asked.

"I read that ghosts like to take things so they have power over the people they stole them from," said Amy, scared.

"We've got to get our things back!" said Ling.

"Why don't we leave a present for the ghost and ask for our things back?" suggested Amy.

"It might work," agreed Ling hopefully.

So, after lunch, the girls made up a small box of presents. Then they asked their mums if they could go on the nature trail again. They were allowed, as long as they were back before dark, and stayed on the path.

Ling's heart was banging like a drum and she felt a lump in her throat. Amy had a flutter in her tummy as she carried the box under her arm. The two girls headed down the path towards the haunted lake.

When they got to the lakeside it was very quiet.

"Let's go back," said Ling.

"We'll just leave the box and run," decided Amy.

They went to the edge of the water and Amy put down the box. But before they could run, Ling cried out in horror.

"GHOST!" she screamed, pointing at the boy standing close by.

The boy had dark, curly hair and Amy knew he was the ghost boy she had seen in the leaflet. The boy jumped in shock and shouted, "Ghost? Where?"

Ling and Amy jumped and fell into the shallow water with a splash.

"Ghost boy. We ... we've brought you a present, so you will stop stealing our things," stammered Amy, still sitting in the water. Amy didn't know where the words came from, but she felt very brave.

"Can I have my hair band back please?" cried Ling.

"Why would I want a hair band?" said the boy rudely, as if Ling were stupid.

"Because ghosts steal things," said Amy. "We know you've taken my sock and my beads, and Ling's watch as well!"

The boy looked puzzled. "I'm not a ghost," he said, "and I don't steal things."

Amy and Ling stood up, cold and wet.

"But you must be a ghost. Your picture is here," said Amy, fishing a wet leaflet from her pocket.

"Oh that!" said the boy. "My dad owns the holiday camp and writes the tourist leaflets. He wrote a story, which he'd heard long ago, to bring more people to the holiday camp. I pretended to be the ghost boy for the picture." The boy gave a cheeky grin.

"Then how were you walking on the water?" said Ling, pointing to the lake.

"We saw you!" said Amy quickly.

As the water lapped across the pebbly shore, the boy raced around the girls and hopped onto something. "Hah, easy trick. I was stone skipping," laughed the boy.

It looked as if the boy were walking on the water. But he was hopping across rocks, hidden just under the water.

"Look," he said proudly. "You can hop right out."

"Cool!" said Amy, stepping out onto a rock, too.

"But who stole our things then?" Ling called out.

The boy skipped back to the shore.

"Well it doesn't sound like anyone in the holiday camp would want a sock and a hair band. I think I know what has happened," he said, grinning.

"What?" asked Ling and Amy at the same time.

"Come on," said the boy. "Time to go on a squirrel hunt."

Chapter 4
A Squirrel Hunt

The boy dived into some nearby bushes, closely followed by Amy and Ling.

"Why a squirrel hunt?" said Ling excitedly, as they crawled along.

"Because squirrels like to steal things," explained the boy. "And this holiday camp is full of squirrels! My name is Matt by the way."

Ling and Amy followed Matt, out of the bushes and over to some trees, close to their own tents.

"The squirrels are so used to people coming to stay in the woods, they sneak into tents all the time," said Matt.

Matt pointed to a nest in a tree and began searching the ground.

"Look," he said, kicking the grass with his trainer.

Something stripy was hidden around the base of the tree.

"My sock!" cried Amy. She picked up the sock, which was wet and slightly chewed around the edges.

"Yuck! I don't think I'll be wearing that again!"

"Can you help me find my watch and hair band, Matt?" asked Ling.

"Sure! We'll find your watch and lots more," he replied proudly.

Ling and Amy looked at each other and laughed. "As long as we don't find a ghost," they both said.